Birkhäuser Advanced Texts
Basler Lehrbücher

Edited by
Herbert Amann, Zürich University

Michel Chipot
ℓ goes to plus infinity

Springer Basel AG

Author:
Michel Chipot
Institut für Mathematik
Abt. Angewandte Mathematik
Universität Zürich
Winterthurerstr. 190
8057 Zürich
Switzerland

2000 Mathematics Subject Classification 35B40; 35J60, 35J65, 35K05, 49J40, 74B05

A CIP catalogue record for this book is available from the
Library of Congress, Washington D.C., USA

Deutsche Bibliothek Cataloging-in-Publication Data
Chipot, Michel:
l goes to plus infinity / Michel Chipot. - Basel ; Boston ; Berlin :
Birkhäuser, 2002
(Birkhäuser advanced texts)
ISBN 978-3-0348-9465-4   ISBN 978-3-0348-8173-9 (eBook)
DOI 10.1007/978-3-0348-8173-9

ISBN 978-3-0348-9465-4

© 2002 Springer Basel AG
Originally published by Birkhäuser Verlag, Basel - Boston - Berlin in 2002
Softcover reprint of the hardcover 1st edition 2002
Member of the BertelsmannSpringer Publishing Group
Printed on acid-free paper produced from chlorine-free pulp. TCF ∞

ISBN 978-3-0348-9465-4

9 8 7 6 5 4 3 2 1   www.birhauser-science.com

Contents

Preface

Many physical problems take place in a cylindrical domain. Under some symmetry condition, when the size ℓ of the cylinder goes to plus infinity, it is reasonable to expect the solution of the problem to be the same in all sections of the cylinder. However, this is not always easy to prove and is almost never established in the literature. We try to fill the gap in this book, in various cases. This gives us the opportunity to introduce a large class of problems with a further goal which should render them more quickly familiar. This issue is also a key point in numerical analysis, when one wants to avoid heavy computations in three dimensions.

We start with an introduction to linear elliptic problems. There we are rather brief. In particular we do not spend too much time on Sobolev spaces, referring the reader to different books. Nevertheless, we introduce Dirichlet, Neumann and mixed problems in a unified way, which is perhaps not always available in the literature.

In the second chapter, we concentrate on the ideas driving the rest of the book. These are explained in very simple situations, and the chapter will give the reader a very good introduction to the basics of the issues developed subsequently.

Chapter 3 develops a general asymptotic theory for linear elliptic equations. Here the problems are addressed in full generality, involving some technical details which will nonetheless be easily accessible with the material of Chapter 2.

In the following chapter we introduce some nonlinear problems from the point of view of existence and uniqueness. First we present a short theory of variational inequalities in $H^1(\Omega)$. Even if most of the results are well known, the presentation here is original. Then, we consider some quasilinear problems, again from the point of view of existence and uniqueness. We end the chapter with a remark on uniqueness for strongly nonlinear problems.

Chapter 5 addresses the question of the asymptotic behaviour of the nonlinear problems introduced in Chapter 4. In contrast with the linear theory, uniqueness could fail for the solution to nonlinear problems. In other words, the limit solution may not be unique, and this creates an additional difficulty. We consider here only cases where the limit solution is unique. In the case of quasilinear problems we will see that a suitable space for obtaining convergence is L^1. Under some additional assumptions we will establish convergence in H^1.

Chapter 6 is an introduction to elliptic systems. We consider first linear systems with a special emphasis on the system of elasticity whose applications are fundamental. We give in particular detailed proofs of the different Korn inequalities which play an essential rôle for linear elasticity. The chapter ends with some existence results for linear and nonlinear systems.

Chapter 7 is devoted to the asymptotic behaviour of elliptic systems. First a linear theory is established. Then special attention is given to the system of elasticity.

In Chapter 8 we provide the reader with an introduction to parabolic problems. We are brief regarding linear theory which is available in many places in the literature. For quasilinear problems we give existence and uniqueness results.

The last chapter contains the asymptotic theory of linear and nonlinear parabolic problems.

Many results presented here are original and have not been published elsewhere. We hope that they can inspire the reader to his own applications.

The book owes much to the discussions that I have had with A. Rougirel. He corrected the proofs with M. Siegwart and I am extremely grateful to both of them for their help. The manuscript has been beautifully typed by Mrs. Schacher and I thank her also very much for her work. Finally, I would like to thank the Swiss National Science Foundation who supported this project under the contract #20-58856.99.

Zürich, December 2000

Chapter 1

Introduction to Linear Elliptic Problems

The basic concepts underlying the topic of linear elliptic problems are well known, thus we will restrict ourselves to some keypoints and some aspects that are not found frequently in the literature. For complementary material we refer the reader to [6], [8], [17], [20], [36], [14].

1.1. The Lax–Milgram theorem

The Lax–Milgram theorem provides an abstract setting for elliptic problems. Let us consider H a real Hilbert space. When no confusion occurs we will denote by $(\cdot, \cdot)$ the scalar product in H and by $|\cdot|$ the associated norm. Let $a = a(u, v)$ be a bilinear, continuous, coercive form on H – i.e., a is a bilinear form on H such that

$$\exists \Lambda \in \mathbb{R} \quad \text{such that} \quad |a(u,v)| \leq \Lambda |u|\,|v| \quad \forall u, v \in H \quad \text{(continuity)}, \tag{1.1}$$

$$\exists \lambda > 0 \quad \text{such that} \quad a(u,u) \geq \lambda |u|^2 \quad \forall u \in H \qquad \text{(coerciveness)}. \tag{1.2}$$

(In the above definition $|\cdot|$ denotes the norm as well as the absolute value.) Let $f \in H'$, where H' denotes the dual of H. Then we can show

THEOREM 1.1 (Lax–Milgram). *Under the above assumptions there exists a unique u solution to*

$$\begin{cases} u \in H, \\ a(u, v) = \langle f, v \rangle \quad \forall v \in H. \end{cases} \tag{1.3}$$

($\langle \cdot \rangle$ denotes the duality bracket between H' and H.) Moreover, if a is symmetric, i.e., if

$$a(u, v) = a(v, u) \quad \forall u, v \in H, \tag{1.4}$$

then u is the unique minimizer of the functional

$$J(v) = \frac{1}{2} a(v, v) - \langle f, v \rangle \tag{1.5}$$

on the space H.

PROOF. For references on Hilbert spaces and topology see for instance [34]. For any $u \in H$ it results from (1.1) that the mapping

$$v \mapsto a(u, v)$$

is a linear continuous form of H – that is to say an element of H'. Set

$$K = \{ v \mapsto a(u, v) \mid u \in H \}. \tag{1.6}$$

It is easy to see that K is a subspace of H'. Moreover we claim:

- K is closed.

Indeed, consider a converging sequence $a(u_n, \cdot)$ of K. Denote by $|\cdot|_*$ the strong dual norm of H'. Since $a(u_n, \cdot)$ is converging this is a Cauchy sequence in H'. One deduces from (1.2) that

$$\lambda|u_n - u_m|^2 \leq a(u_n - u_m, u_n - u_m) \leq |a(u_n, \cdot) - a(u_m, \cdot)|_* |u_n - u_m|$$

$$\Rightarrow \quad |u_n - u_m| \leq \frac{1}{\lambda}|a(u_n, \cdot) - a(u_m, \cdot)|_*, \tag{1.7}$$

i.e., u_n is a Cauchy sequence in H. Since H is complete this sequence converges toward an element that we will denote u. Then it holds that

$$|a(u_n, v) - a(u, v)| = |a(u_n - u, v)| \leq \Lambda|u_n - u| \, |v|$$

$$\Rightarrow \quad |a(u_n, \cdot) - a(u, \cdot)|_* \leq \Lambda|u_n - u| \tag{1.8}$$

and K is closed since $a(u_n, \cdot)$ converges toward $a(u, \cdot)$. We claim now that

- K is dense in H'.

Indeed, let v be in the orthogonal of K. It holds that

$$a(u, v) = 0 \quad \forall u \in H$$

$$\Rightarrow \quad \lambda|v|^2 \leq a(v, v) = 0 \quad \Rightarrow \quad v = 0. \tag{1.9}$$

K being dense and closed, $K = H'$ and the existence of u solution to (1.3) follows. To show uniqueness – we notice that if u' is another solution it holds that

$$a(u, v) = a(u', v) \quad \forall v \in H$$

$$\Rightarrow \quad 0 = a(u - u', u - u') \geq \lambda|u - u'|^2 \tag{1.10}$$

and uniqueness follows.

To show now that u minimizes (1.5) it is enough to notice that

$$J(v) = J(u + v - u) = \frac{1}{2}a(u + v - u, u + v - u) - \langle f, v \rangle$$

$$= \frac{1}{2}a(u, u) + a(u, v - u) + \frac{1}{2}a(v - u, v - u) - \langle f, v \rangle$$

$$= \frac{1}{2}a(u, u) + \langle f, v - u \rangle + \frac{1}{2}a(v - u, v - u) - \langle f, v \rangle$$

$$= J(u) + \frac{1}{2}a(v - u, v - u) > J(u) \quad \forall v \neq u. \tag{1.11}$$

This completes the proof of the theorem. $\square$

1.2. Elementary notions on Sobolev spaces

We recall here only the basic notions or some properties less available in the literature. For complements the reader can consult the references at the beginning of this chapter.

Let Ω denote a bounded open subset of $\mathbb{R}^n$, $n \geq 1$, that we will assume later to be Lipschitz continuous. Then

$$H^1(\Omega) = \{\, v \in L^2(\Omega) \mid \partial_{x_i} v \in L^2(\Omega) \ \forall\, i = 1, \ldots, n \,\}. \qquad (1.12)$$

Here $L^2(\Omega)$ denotes the space of the "class" of square integrable functions – in short

$$L^2(\Omega) = \left\{\, v : \Omega \to \mathbb{R} \text{ measurable} \,\middle|\, \int_\Omega v^2(x)\,dx < +\infty \,\right\}. \qquad (1.13)$$

∂_{x_i} denotes the derivative in the direction x_i. In (1.12) it is taken in the distributional or weak sense (see [17], [29], [20]).

Recall that $L^2(\Omega)$ is a Hilbert space for the scalar product

$$(u, v)_{2,\Omega} = \int_\Omega u(x)v(x)\,dx \qquad (1.14)$$

and the norm

$$|u|_{2,\Omega} = \left\{\, \int_\Omega u(x)^2\,dx \,\right\}^{\frac{1}{2}}. \qquad (1.15)$$

One can show easily that $H^1(\Omega)$ is also a Hilbert space when equipped with the scalar product

$$(u, v)_{1,2} = \int_\Omega \nabla u \cdot \nabla v + u \cdot v\,dx. \qquad (1.16)$$

(∇u denotes the vector $(\partial_{x_1} u, \partial_{x_2} u, \ldots, \partial_{x_n} u)$.)

The associated norm is – if $|\nabla u|$ denotes the euclidean norm of ∇u –

$$|u|_{1,2} = \left\{\, \int_\Omega |\nabla u|^2 + u^2\,dx \,\right\}^{\frac{1}{2}}. \qquad (1.17)$$

For properties on $H^1(\Omega)$ we refer the reader to [17], [19], [20], [6], [8]. One remarkable property is

THEOREM 1.2. *Assume that Ω is a bounded Lipschitz domain of $\mathbb{R}^n$; then the canonical embedding from $H^1(\Omega)$ into $L^2(\Omega)$ is compact.*

PROOF. See [19], [29]. □

The use of Theorem 1.2 in partial differential equations is the following: from any bounded sequence in $H^1(\Omega)$ we can extract a subsequence converging weakly in $H^1(\Omega)$ and strongly in $L^2(\Omega)$. We will see later some applications of this principle.

Also let us now introduce a class of equivalent norms on $H^1(\Omega)$. For that purpose let us denote by a a function in $L^\infty(\Omega)$ such that

$$0 \leq a \leq A \quad \text{a.e. in } \Omega, \quad a \not\equiv 0. \qquad (1.18)$$

Then, let us define

$$|u|_a = \left\{ \int_\Omega |\nabla u|^2 + au^2 \, dx \right\}^{\frac{1}{2}}. \qquad (1.19)$$

Then we have (see [24])

THEOREM 1.3. *On $H^1(\Omega)$ the norm $|\cdot|_a$ and the classical norm $|\cdot|_{1,2}$ are equivalent. That is to say there exist two constants c_1, c_2 such that*

$$c_1|u|_a \leq |u|_{1,2} \leq c_2|u|_a \quad \forall u \in H^1(\Omega). \qquad (1.20)$$

(Note that the norm (1.19) derives from a scalar product).

PROOF. From (1.18) we derive

$$|u|_a^2 = \int_\Omega |\nabla u|^2 + au^2 \, dx \leq \text{Max}(1, A) \int_\Omega |\nabla u|^2 + u^2 \, dx. \qquad (1.21)$$

This proves the left-hand side inequality of (1.20). For the right-hand side, suppose that it fails. Then, for every n there exists $u_n \in H^1(\Omega)$ such that

$$|u_n|_{1,2} \geq n|u_n|_a.$$

Considering $v_n = u_n/|u_n|_{1,2}$ we have exhibited a sequence v_n such that

$$|v_n|_a \leq \frac{1}{n}, \qquad |v_n|_{1,2} = 1. \qquad (1.22)$$

From Theorem 1.2 it follows that – up to a subsequence – we can assume that there exists a $v \in H^1(\Omega)$ such that

$$v_n \rightharpoonup v \quad \text{in} \quad H^1(\Omega), \qquad (1.23)$$

$$v_n \to v \quad \text{in} \quad L^2(\Omega). \qquad (1.24)$$

In particular we have for every $i = 1, \ldots, n$

$$\partial_{x_i} v_n \to \partial_{x_i} v \quad \text{in} \quad \mathcal{D}'(\Omega).$$

$\mathcal{D}'(\Omega)$ denotes the space of distributions on Ω – see [33], [17]. But from (1.22) we have

$$\partial_{x_i} v_n \to 0 \quad \text{in} \quad L^2(\Omega) \quad \Rightarrow \quad \nabla v = 0 \quad \Rightarrow \quad v = \text{cst}.$$

Moreover, exploiting again (1.22) one has

$$|\sqrt{a} v_n|_{2,\Omega}^2 = \int_\Omega a v_n^2 \, dx \to 0. \qquad (1.25)$$

and also

$$\left| |\sqrt{a} v_n|_{2,\Omega} - |\sqrt{a} v|_{2,\Omega} \right| \leq |\sqrt{a}(v_n - v)|_{2,\Omega} \leq \sqrt{A}|(v_n - v)|_{2,\Omega} \to 0.$$

This implies

$$\int_\Omega a v^2 \, dx = 0 \quad \Rightarrow \quad v = 0$$

by (1.18). Thus, we have obtained

$$\nabla v_n \to 0 \quad \text{in} \quad L^2(\Omega), \qquad v_n \to 0 \quad \text{in} \quad L^2(\Omega)$$

which renders $|v_n|_{1,2} = 1$ impossible. Note that if we use (1.24) locally no assumption on Ω is necessary. This completes the proof of the theorem.   □

An important subspace of $H^1(\Omega)$ is $H_0^1(\Omega)$ defined as

$$H_0^1(\Omega) = \text{ the closure of } \mathcal{D}(\Omega) \text{ in } H^1(\Omega) \tag{1.26}$$

$$= \{\, v \in H^1(\Omega) \mid v = 0 \text{ on } \Gamma \,\}. \tag{1.27}$$

($\mathcal{D}(\Omega)$ denotes the space of infinitely differentiable functions with compact support in Ω, $\Gamma = \partial\Omega$ the boundary of Ω.) $H_0^1(\Omega)$ is the space of functions of $H^1(\Omega)$ vanishing on the boundary of Ω in a certain sense. The sense adopted in (1.26) is clear, the one in (1.27) is in the sense of the trace. This imposes some regularity on $\partial\Omega$ – we refer the reader to [6], [8] for details. Let us denote by Γ_0 a subset of Γ of positive measure – i.e.,

$$|\Gamma_0| \neq 0.$$

($|\cdot|$ denotes the superficial measure on Γ – this supposes that Γ is a Lipschitz boundary which we will assume.) Define $C_0^1(\overline{\Omega}, \Gamma_0)$ as the set of functions continuously differentiable in $\overline{\Omega}$ and vanishing on Γ_0. Then, let us denote by V the space

$$V = \text{ the closure of } C_0^1(\overline{\Omega}; \Gamma_0) \text{ in } H^1(\Omega) \tag{1.28}$$

(in short V is the set of functions of $H^1(\Omega)$ vanishing – in a certain sense – on Γ_0.) V is of course a closed subspace of $H^1(\Omega)$. We will set

$$|v|_V = \left\{ \int_\Omega |\nabla v(x)|^2 \, dx \right\}^{\frac{1}{2}} = \left\| \nabla v \right\|_{2,\Omega}. \tag{1.29}$$

Then, we have:

THEOREM 1.4. *On V the norms (1.29), (1.17) are equivalent in the sense that there exist constants c_1, c_2 such that*

$$c_1|v|_V \leq |v|_{1,2} \leq c_2|v|_V \quad \forall v \in V. \tag{1.30}$$

PROOF. The inequality of the left-hand side of (1.30) holds clearly with $c_1 = 1$. To prove the other inequality we will argue by contradiction. Suppose that the inequality of the right-hand side of (1.30) fails. Then, for any $n \in \mathbb{N}$, there exists $u_n \in V$ such that we have

$$|u_n|_{1,2} \geq n|u_n|_V. \tag{1.31}$$

Setting $v_n = u_n/|u_n|_{1,2}$ it holds then for v_n that

$$|v_n|_V \leq \frac{1}{n}, \qquad |v_n|_{1,2} = 1. \tag{1.32}$$

From Theorem 1.2 – up to a subsequence – we can suppose that for some $v \in V$ we have

$$v_n \rightharpoonup v \text{ in } H^1(\Omega), \qquad v_n \to v \text{ in } L^2(\Omega). \tag{1.33}$$

In particular this implies that for every $i = 1, \ldots, n$

$$\partial_{x_i} v_n \to \partial_{x_i} v \text{ in } \mathcal{D}'(\Omega).$$

Now from (1.32) we derive that

$$\partial_{x_i} v_n \to 0 \quad \text{in} \quad L^2(\Omega)$$

and thus by uniqueness of the limit in $\mathcal{D}'(\Omega)$ it holds that

$$\nabla v = 0 \quad \Rightarrow \quad v = \text{cst} \quad \text{in } \Omega. \tag{1.34}$$

By the definition of V and the definition of the trace (see [6], [29], [8]) one has

$$\gamma_0(v) = c = 0 \quad \text{on} \quad \Gamma_0.$$

($\gamma_0(v)$ denotes the trace of v on Γ.) It follows that

$$v_n \to 0 \quad \text{in} \quad L^2(\Omega)$$

and by (1.32)

$$v_n \to 0 \quad \text{in} \quad H^1(\Omega).$$

This contradicts $|v_n|_{1,2} = 1$ and completes the proof. □

As an immediate consequence we have

THEOREM 1.5 (Poincaré's inequality on V). *There exists a constant c such that*

$$|v|_{2,\Omega} \le c|v|_V = c\|\nabla v\|_{2,\Omega} \quad \forall v \in V. \tag{1.35}$$

PROOF. From (1.30) we derive

$$\int_\Omega |\nabla v|^2 + v^2 \, dx \le c_2^2 \int_\Omega |\nabla v|^2 \, dx$$

and thus

$$\int_\Omega v^2 \, dx \le (c_2^2 - 1) \int_\Omega |\nabla v|^2 \, dx.$$

This completes the proof. □

REMARK 1.1. We will also adopt the notation

$$V = H_0^1(\Omega; \Gamma_0). \tag{1.36}$$

We denote by $H^{-1}(\Omega)$ the dual space of $H_0^1(\Omega)$. We have:

THEOREM 1.6. $H^{-1}(\Omega)$ *is the set of distributions of the type*

$$T = f_0 + \sum_{i=1}^n \partial_{x_i} f_i \tag{1.37}$$

where $f_0, f_1, \ldots, f_n \in L^2(\Omega)$.

PROOF. Let T be a distribution of the type (1.37). It holds that for any $\varphi \in \mathcal{D}(\Omega)$

$$|\langle T, \varphi \rangle| = \left| \int_\Omega \{f_0 \cdot \varphi - \sum_{i=1}^n f_i \partial_{x_i} \varphi\} \, dx \right| \leq ||f||_{2,\Omega} |\varphi|_{1,2}$$

where $|f| = \{\sum_{i=0}^n |f_i|^2\}^{1/2}$ – (this follows from the Cauchy–Schwarz inequality). Thus T is a linear form on $\mathcal{D}(\Omega)$ equipped with the topology of $H_0^1(\Omega)$ and can thus be extended in a unique way into a linear form on $H_0^1(\Omega)$. Conversely, let $T \in H^{-1}(\Omega)$. Then, from the Riesz representation theorem it follows that there exists a unique $u \in H_0^1(\Omega)$ such that

$$\int_\Omega \nabla u \cdot \nabla v + u \cdot v \, dx = \langle T, v \rangle \quad \forall v \in H_0^1(\Omega).$$

Setting $f_0 = u$, $f_i = -\partial_{x_i} u$, $i = 1, \ldots, n$ it follows that (1.37) holds. This completes the proof of the theorem. □

Let us denote by Ω' a subset of Ω of positive measure. Let us set

$$W = W(\Omega; \Omega') = \left\{v \in H^1(\Omega) \,\middle|\, \int_{\Omega'} v \, dx = 0\right\}. \tag{1.38}$$

Then, clearly, W is a closed subspace of $H^1(\Omega)$ and in the same spirit as above it holds that:

THEOREM 1.7. *There exist two constants c_1, c_2 such that*

$$c_1 ||\nabla v||_{2,\Omega} \leq |v|_{1,2} \leq c_2 ||\nabla v||_{2,\Omega} \quad \forall v \in W. \tag{1.39}$$

Moreover, there is a constant c such that

$$|v|_{2,\Omega} \leq c ||\nabla v||_{2,\Omega} \quad \forall v \in W. \tag{1.40}$$

PROOF. The Poincaré inequality (1.40) is derived from (1.39) exactly as in Theorem 1.5. The inequality of the left-hand side of (1.39) holds for $c_1 = 1$. To prove the other inequality we proceed exactly as in the proof of Theorem 1.4. To show that $v = 0$ we remark that arriving at (1.34) – since W is closed and thus weakly closed in $H_0^1(\Omega)$ – we have $v = cst$, $v \in W$. This implies by the definition of W that $v = 0$ and we conclude as in the proof of theorem 1.4. This completes the proof of the Theorem. □

We will also use higher order Sobolev spaces. For instance

$$H^k(\Omega) = \{v \in L^2(\Omega) \mid D^\alpha v \in L^2(\Omega) \, \forall \alpha, |\alpha| \leq k\}$$

$(\alpha = (\alpha_1, \ldots, \alpha_n) \in \mathbb{N}^n, |\alpha| = \alpha_1 + \cdots + \alpha_n, D^\alpha v = \partial_{x_1}^{\alpha_1} \partial_{x_2}^{\alpha_2} \ldots \partial_{x_n}^{\alpha_n} v$, the derivative being taken in the distributional sense – we refer to [6], [17], [20] for details). These spaces are Hilbert spaces for the Hilbert norm

$$|v|_{k,2} = \left\{\sum_{|\alpha| \leq k} |D^\alpha v|_{2,\Omega}^2\right\}^{1/2}.$$

1.3. Applications to linear elliptic problems

In this section we apply the results of the preceding sections to solve in a weak sense problems of the elliptic type.

Let Ω be a bounded open subset of $\mathbb{R}^n$ with a Lipschitz boundary Γ. Let a and a_{ij}, $i, j = 1, \ldots, n$, be functions in $L^\infty(\Omega)$ such that for $\lambda > 0$ it holds that:

$$(\text{ellipticity}) \qquad a_{ij}(x)\xi_i\xi_j \geq \lambda|\xi|^2 \quad \text{a.e. } x \in \Omega, \; \forall \xi \in \mathbb{R}^n. \tag{1.41}$$

In the above formula we make the Einstein convention that repeated indices mean a summation from 1 to n. On a we suppose

$$a(x) \geq 0 \quad \text{a.e. } x \in \Omega. \tag{1.42}$$

Then, for $u, v \in H^1(\Omega)$, we define

$$a(u, v) = \int_\Omega a_{ij}(x)\partial_{x_j}u\partial_{x_i}v + a(x)uv \, dx. \tag{1.43}$$

Note that since a_{ij}, $a \in L^\infty(\Omega)$ the above definition makes sense. Moreover we can state

THEOREM 1.8. *Under the above assumptions* $a(\cdot, \cdot)$ *is a continuous bilinear form on* $H^1(\Omega)$. *Moreover,* $a(\cdot, \cdot)$ *is coercive on* $H_0^1(\Omega)$, $H_0^1(\Omega; \Gamma_0)$, W. *If* $a \not\equiv 0$, $a(\cdot, \cdot)$ *is also coercive on* $H^1(\Omega)$.

PROOF. The bilinearity is clear. Since a_{ij}, $a \in L^\infty$ one has for some constant c

$$|a(u, v)| \leq \int_\Omega c|\nabla u|\,|\nabla v| + c|u|\,|v|\, dx \leq c|u|_{1,2}|v|_{1,2} \quad \forall u, v \in H^1(\Omega). \tag{1.44}$$

This shows the continuity of a for the $H^1(\Omega)$-topology.

Next, we notice that (1.41) implies that

$$a(u, u) \geq \lambda \int_\Omega |\nabla u|^2 \, dx = \lambda||\nabla u||^2_{2,\Omega}. \tag{1.45}$$

The coerciveness of a for the $H^1(\Omega)$-topology on $H_0^1(\Omega) = H_0^1(\Omega; \Gamma)$, $H_0^1(\Omega; \Gamma_0)$, W follows then from the Theorems 1.4, 1.7. In the case where $a \not\equiv 0$,

$$a(u, u) \geq \lambda \int_\Omega |\nabla u|^2 + \int_\Omega au^2 \, dx \geq \min(\lambda, 1)|u|^2_a \tag{1.46}$$

and the coerciveness follows from Theorem 1.3. This completes the proof. $\qquad\square$

Then we can show

THEOREM 1.9 (Dirichlet Problem). *Under the above assumptions – i.e., when* (1.41), (1.42) *hold for any* $f \in H^{-1}(\Omega)$, *there exists a unique* u *solution to*

$$\begin{cases} a(u, v) = \langle f, v \rangle \quad \forall v \in H_0^1(\Omega), \\ u \in H_0^1(\Omega). \end{cases} \tag{1.47}$$

PROOF. This is an immediate consequence of Theorem 1.1 and Theorem 1.8. $\qquad\square$

REMARK 1.2. One says that u is the weak solution to the Dirichlet problem

$$\begin{cases} -\partial_{x_i}(a_{ij}(x)\partial_{x_j}u) + au = f & \text{in } \Omega, \\ u = 0 \text{ on } \Gamma. \end{cases} \tag{1.48}$$

Note that in this case a could vanish identically.

THEOREM 1.10 (Mixed boundary conditions problem). *Let $V = H_0^1(\Omega; \Gamma_0)$ with $|\Gamma_0| > 0$. Then, for every $f \in V'$ there exists a unique u solution to*

$$\begin{cases} a(u,v) = \langle f, v \rangle & \forall v \in V, \\ u \in V. \end{cases} \tag{1.49}$$

PROOF. Immediate by the conjunction of Theorems 1.1 and 1.8. □

REMARK 1.3. u is called the weak solution to the mixed Dirichlet-Neumann boundary condition problems. Suppose for instance that f is a function of $L^2(\Omega)$. It is clear that it holds in the distributional sense that

$$-\partial_{x_i}(a_{ij}(x)\partial_{x_j}u) + au = f \quad \text{in } \Omega. \tag{1.50}$$

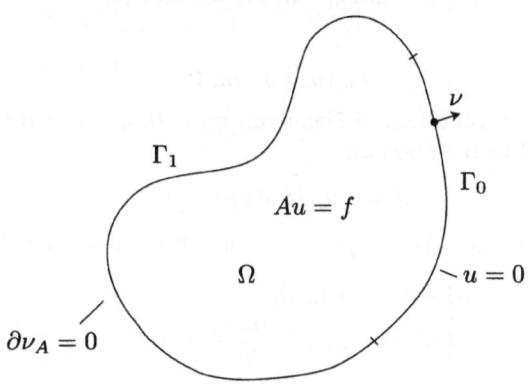

FIGURE 1.1

Moreover – in a weak sense – one has $u = 0$ on Γ_0. Suppose now that all the data of the problem are smooth. Then, for v smooth, $v \in H_0^1(\Omega; \Gamma_0)$ it holds that

$$\int_\Omega a_{ij}(x)\partial_{x_j}u\partial_{x_i}v + auv\, dx = \langle f, v \rangle$$

i.e.,

$$\int_\Omega \partial_{x_i}(a_{ij}\partial_{x_j}uv) + \{-\partial_{x_i}(a_{ij}\partial_{x_j}u) + au\}v\, dx = \langle f, v \rangle.$$

Now, if all the data are smooth, (1.50) holds in the usual sense and

$$\int_\Omega \partial_{x_i}(a_{ij}\partial_{x_j}uv)\, dx = 0. \tag{1.51}$$

Applying the divergence theorem and denoting by ν the outward unit normal to Γ we derive

$$\int_{\Gamma} a_{ij} \partial_{x_j} u \nu_i v \, d\sigma = 0. \tag{1.52}$$

(We refer for instance to [8] for the divergence formula. $\nu = (\nu_1, \ldots, \nu_n)$, $d\sigma$ denotes the superficial measure on Γ.)

Since $v = 0$ on Γ_0, if $\Gamma_1 = \Gamma \setminus \Gamma_0$, then

$$\int_{\Gamma_1} a_{ij} \partial_{x_j} u \nu_i v \, d\sigma = 0.$$

Since v is arbitrary it follows easily – again in the case of smooth data – that

$$\partial \nu_A u \stackrel{\text{Def}}{=} a_{ij} \partial_{x_j} u \nu_i = 0 \quad \text{on } \Gamma_1.$$

To summarize (see also Figure 1.1) u is the weak solution to

$$\begin{cases} -\partial_{x_i}(a_{ij}(x)\partial_{x_j}u) + au = f & \text{in } \Omega, \\ u = 0 \text{ on } \Gamma_0, \quad \partial \nu_A u = 0 \text{ on } \Gamma_1. \end{cases} \tag{1.53}$$

The condition

$$\partial \nu_A u = 0 \quad \text{on } \Gamma_1 \tag{1.54}$$

is called a boundary condition of Neumann type. $\partial \nu_A$ is called the conormal derivative associated to the operator

$$A = -\partial_{x_i}(a_{ij}(x)\partial_{x_j}) + a. \tag{1.55}$$

For instance in the case where $a_{ij}(x) = \delta_{ij}$, $a = 0$ we have solved the problem

$$\begin{cases} -\Delta u = f \text{ in } \Omega, \\ u = 0 \text{ on } \Gamma_0, \quad \dfrac{\partial u}{\partial \nu} = 0 \text{ on } \Gamma_1, \end{cases} \tag{1.56}$$

in a weak sense. ($\frac{\partial u}{\partial \nu} = \nabla u \cdot \nu$ is the normal derivative to Γ.) This is the Laplace equation with mixed Dirichlet–Neumann boundary conditions. In the case where $\Gamma_0 = \emptyset$ we have

THEOREM 1.11 (Neumann Problem). *Under the above conditions – i.e., when* (1.41), (1.42) *holds and if in addition*

$$a \not\equiv 0 - i.e., \ a > 0 \ on \ a \ set \ of \ positive \ measure, \tag{1.57}$$

then, for every $f \in (H^1(\Omega))'$, *dual of* $H^1(\Omega)$, *there exists a unique* u *solution to*

$$\begin{cases} a(u, v) = \langle f, v \rangle \quad \forall v \in H^1(\Omega), \\ u \in H^1(\Omega). \end{cases} \tag{1.58}$$

PROOF. The proof results immediately from Theorems 1.1 and 1.8. □

REMARK 1.4. As in Remark 1.3 we can show when f is a function that u is a weak solution to the problem

$$\begin{cases} -\partial_{x_i}(a_{ij}(x)\partial_{x_j}u) + au = f & \text{in } \Omega, \\ \partial\nu_A u = 0 \text{ on } \Gamma. \end{cases} \tag{1.59}$$

Note that the first equation of (1.59) holds in the distributional sense since a continuous linear form on $H^1(\Omega)$ restricted to $\mathcal{D}(\Omega)$ is a distribution. However, two distinct linear forms can have the same restrictions so that $H^1(\Omega)'$ is not identifiable to a space of distributions. This will be clearer in the proposition below.

If Ω is a bounded open set and $\alpha \in (0,1)$ we denote by $C^\alpha(\overline{\Omega})$ the space

$$C^\alpha(\overline{\Omega}) = \{ v : \overline{\Omega} \to \mathbb{R}, v \text{ continuous} \mid \exists\, C \text{ such that} $$
$$|v(x) - v(y)| \le C|x - y|^\alpha \ \forall x, y \in \Omega \}. \tag{1.60}$$

$C^\alpha(\overline{\Omega})$ is the space of Hölder continuous functions on Ω with exponent α. It is easy to see that $C^\alpha(\overline{\Omega})$ is a Banach space when equipped with the norm

$$|v|_{C^\alpha(\overline{\Omega})} = |v|_\infty + \operatorname*{Sup}_{\substack{x,y\in\Omega \\ x\ne y}} \frac{|v(x) - v(y)|}{|x - y|^\alpha}. \tag{1.61}$$

($|\ |_\infty$ denotes the usual L^∞-norm.) In dimension 1 we have

PROPOSITION 1.12. *Let $\Omega = (-1, 1)$. It holds that*

$$H^1(\Omega) \hookrightarrow C^{\frac{1}{2}}(\overline{\Omega}). \tag{1.62}$$

($\hookrightarrow$ means that the canonical embedding is continuous.) Moreover, every element $T \in H^1(\Omega)'$ can be written as

$$T = \alpha\delta_1 + \beta\delta_{-1} + T_f \tag{1.63}$$

where δ_1, δ_{-1} denote the Dirac masses at 1, -1 and T_f is a continuous linear form defined on $H^1(\Omega)$ by

$$\langle T_f, v \rangle = (f_0, v)_{2,\Omega} + (f_1, v')_{2,\Omega} \tag{1.64}$$

where $f_0, f_1 \in L^2(\Omega)$.

PROOF. Let v be a smooth function of $H^1(\Omega)$. It holds that

$$|v(x) - v(y)| = \left| \int_x^y v'(s)\, ds \right| \le \int_x^y |v'(s)|\, ds \le |v'|_{2,\Omega}|x - y|^{\frac{1}{2}} \tag{1.65}$$

by the Cauchy–Schwarz inequality. Moreover, by the intermediate value theorem there exists $y \in \Omega$ such that

$$v(y) = \frac{1}{|\Omega|} \int_\Omega v(s)\, ds = \frac{1}{2} \int_\Omega v(s)\, ds. \tag{1.66}$$

($|\Omega|$ denotes the measure of Ω.)

Thus, by (1.65), we obtain for every $x \in \Omega$ – again using the Cauchy–Schwarz inequality

$$|v(x)| \leq |v(x) - v(y)| + |v(y)|$$

$$\leq \sqrt{2}|v'|_{2,\Omega} + \frac{1}{2}\left|\int_\Omega v(s)\,ds\right| \leq \sqrt{2}|v'|_{2,\Omega} + \frac{1}{\sqrt{2}}|v|_{2,\Omega}. \tag{1.67}$$

Combining (1.61), (1.65), (1.67) we derive easily

$$|v|_{C^{1/2}(\overline{\Omega})} \leq C\{|v'|_{2,\Omega} + |v|_{2,\Omega}\} \leq C|v|_{1,2}. \tag{1.68}$$

Since this is true for instance for any C^1 function on $\overline{\Omega}$, (1.62) follows by density of these smooth functions in $H^1(\Omega)$. For these questions of density see for instance [17], [19], [1].

From (1.62) it results that (1.63) defines a continuous linear form on $H^1(\Omega)$. Conversely, let $T \in H^1(\Omega)'$. Consider θ_1, θ_{-1} two smooth functions such that

$$\theta_1(1) = \theta_{-1}(-1) = 1, \qquad \theta_1(-1) = \theta_{-1}(1) = 0.$$

Then we have, since $v = \theta_1 v(1) + \theta_{-1} v(-1) + v - \theta_1 v(1) - \theta_{-1} v(-1)$

$$\langle T, v \rangle = \langle T, \theta_1 \rangle v(1) + \langle T, \theta_{-1} \rangle v(-1) + \langle \tilde{T}, v - \theta_1 v(1) - \theta_{-1} v(-1) \rangle \tag{1.69}$$

where $\tilde{T}$ denotes the restriction of T to $H_0^1(\Omega)$. Since $\tilde{T}$ is a linear continuous form on $H_0^1(\Omega)$ it follows from Theorem 1.6 that

$$\langle \tilde{T}, v - \theta_1 v(1) - \theta_{-1} v(-1) \rangle$$

$$= \int_\Omega f_0\{v - \theta_1 v(1) - \theta_{-1} v(-1)\} - f_1\{v' - \theta_1' v(1) - \theta_{-1}' v(-1)\}\,dx.$$

Combining this with (1.69) we obtain for some α, β

$$\langle T, v \rangle = \alpha v(1) + \beta v(-1) + \int_\Omega f_0 v - f_1 v'\,dx$$

which completes the proof, changing eventually $-f_1$ into f_1.   □

REMARK 1.5. In the Remarks 1.3, 1.4 it was necessary to have $\alpha = \beta = 0$ in (1.63). If $\Omega = (-1, 1)$, suppose that u is the solution to

$$\begin{cases} \int_\Omega u'v' + uv\,dx = \langle T, v \rangle = v(1) \quad \forall v \in H^1(\Omega), \\ u \in H^1(\Omega). \end{cases} \tag{1.70}$$

Taking $v \in H_0^1(\Omega)$ one derives

$$-u'' + u = 0 \quad \Rightarrow \quad u = a_1 e^x + a_2 e^{-x}.$$

Plugging this into (1.70) and integrating by parts,

$$\int_\Omega (u'v)' - u''v + uv \, dx = v(1) \quad \forall v \in H^1(\Omega)$$

$$\Rightarrow \quad u'(1)v(1) - u'(-1)v(-1) = v(1) \quad \forall v \in H^1(\Omega)$$

$$\Rightarrow \quad u'(1) = 1, \quad u'(-1) = 0,$$

which determines u without ambiguity. Note that $u'(1) \neq 0$ in this case.

In the case of pure Neumann boundary conditions we turn now to the case where (1.57) fails, i.e., to the case where $a = 0$. So, in this case we have

$$a(u, v) = \int_\Omega a_{ij}(x) \partial_{x_j} u \partial_{x_i} v \, dx. \tag{1.71}$$

Taking $v = 1$ in (1.58) we see then that existence of a solution to (1.58) in this case will impose

$$\langle f, 1 \rangle = 0. \tag{1.72}$$

But then, if u is a solution, so is $u + C$ for any constant C. So, in order to have uniqueness of a solution we have to impose for instance for some $\Omega' \subset \Omega$

$$\int_{\Omega'} u \, dx$$

to be fixed. With this in mind we have

THEOREM 1.13. *Let Ω' be a set of positive measure in Ω. Let $f \in H^1(\Omega)'$ satisfying (1.72); then there exists a unique $u \in H^1(\Omega)$ solution to*

$$\begin{cases} a(u, v) = \langle f, v \rangle \quad \forall v \in H^1(\Omega), \\ u \in H^1(\Omega), \quad \int_{\Omega'} u \, dx = C, \end{cases} \tag{1.73}$$

where C is a fixed constant. Recall that $a(u,v)$ is given by (1.71).

PROOF. We notice that

$$w = u - C/|\Omega'|$$

satisfies – see the notation (1.38)

$$\begin{cases} a(w, v) = \langle f, v \rangle \quad \forall v \in H^1(\Omega), \\ w \in W. \end{cases} \tag{1.74}$$

Now, it follows from Theorems 1.7, 1.8 that there exists a unique solution w to (1.74). Then $u = w + C/|\Omega'|$ is the unique solution to (1.73). This completes the proof of the theorem. (In the above proof we used the fact that any function of $H^1(\Omega)$ is the sum of a function of W and a constant.) $\qquad \square$

We now briefly investigate the case of some nonhomogeneous boundary conditions. For that we consider

$$g \in H^1(\Omega), \quad h \in L^2(\Gamma), \quad f_0, f_1, \ldots, f_n \in L^2(\Omega) \tag{1.75}$$

when Ω is a Lipschitz open subset of $\mathbb{R}^n$. Due to the trace theorem (see [29], [17], [8]) the mapping

$$v \mapsto \int_\Gamma hv d\sigma(x) + \int_\Omega f_0 v + f_i \partial_{x_i} v \, dx = \langle f, v \rangle \tag{1.76}$$

is a continuous linear form on $H^1(\Omega)$. Then, its restriction to $H_0^1(\Omega)$, $H_0^1(\Omega; \Gamma_0)$ is also continuous and we have

THEOREM 1.14. *Suppose that we are under the assumptions of Theorem 1.8. Let V be either $H_0^1(\Omega)$, $H_0^1(\Omega; \Gamma_0)$ or $H^1(\Omega)$ and f given by (1.75), (1.76). Then, there exists a unique u solution to*

$$\begin{cases} a(u, v) = \langle f, v \rangle & \forall v \in V, \\ u - g \in V. \end{cases} \tag{1.77}$$

PROOF. It is enough to notice that u satisfies (1.77) iff

$$\tilde{u} = u - g$$

satisfies

$$\begin{cases} a(\tilde{u} + g, v) = \langle f, v \rangle & \forall v \in V, \\ \tilde{u} \in V, \end{cases}$$

i.e., $\tilde{u}$ is the unique solution to

$$\begin{cases} a(\tilde{u}, v) = \langle f, v \rangle + a(-g, v) & \forall v \in V, \\ \tilde{u} \in V, \end{cases}$$

which has a unique solution.   □

REMARK 1.6. Clearly u solution to (1.77) is a weak solution to

$$\begin{cases} -\partial_{x_i}(a_{ij}\partial_{x_j}u) + au = f_0 - \partial_{x_i}f_i \text{ in } \Omega, \\ u = g \text{ on } \Gamma_0, \quad \partial_{\nu_A}u = h \text{ on } \Gamma_1. \end{cases} \tag{1.78}$$

(With an obvious meaning for Γ_1.)

We establish now the so-called maximum principle (see also [28]).

THEOREM 1.15. *Suppose that we are under the assumptions of Theorem 1.8 with V either $H_0^1(\Omega)$, $H_0^1(\Omega; \Gamma_0)$ or $H^1(\Omega)$. Let $f_i \in V'$, $i = 1, 2$ such that*

$$\langle f_1, v \rangle \le \langle f_2, v \rangle \quad \forall v \in V, \ v \ge 0. \tag{1.79}$$

Then, if u_i, $i = 1, 2$ denotes the solution to

$$\begin{cases} a(u_i, v) = \langle f_i, v \rangle & \forall v \in V, \\ u_i \in V, \end{cases} \tag{1.80}$$

it holds that

$$u_1 \le u_2 \quad a.e. \text{ in } \Omega. \tag{1.81}$$

PROOF. Let φ be a smooth function such that

$$\varphi(t) = 0 \quad \forall t \leq 0, \qquad \varphi(t) > 0, \quad \forall t > 0, \qquad 0 \leq \varphi'(t) \leq 1 \quad \forall t > 0. \quad (1.82)$$

Suppose that we have proved

LEMMA 1.16. *Let φ be a C^1-function such that*

$$\varphi(0) = 0, \qquad |\varphi'(t)| \leq c \quad (1.83)$$

for some constant c. Let $v \in V$. Then $\varphi(v) \in V$ and it holds that

$$\partial_{x_i}(\varphi(v)) = \varphi'(v)\partial_{x_i}v. \quad (1.84)$$

Then, from (1.80) we derive

$$a(u_1 - u_2, v) = \langle f_1 - f_2, v \rangle \quad \forall v \in V. \quad (1.85)$$

Taking $v = \varphi(u_1 - u_2)$ in (1.85) it follows by (1.79) that

$$a(u_1 - u_2, \varphi(u_1 - u_2)) = \langle f_1 - f_2, \varphi(u_1 - u_2) \rangle \leq 0.$$

But, from the Lemma, we derive

$$\int_\Omega a_{ij}\partial_{x_j}(u_1 - u_2)\partial_{x_i}(u_1 - u_2)\varphi'(u_1 - u_2)$$
$$+ a(u_1 - u_2)\varphi(u_1 - u_2)\,dx \leq 0. \quad (1.86)$$

It follows – due to the ellipticity conditions and (1.82) – that

$$|\nabla(u_1 - u_2)|^2\varphi'(u_1 - u_2), \; a\varphi(u_1 - u_2) = 0 \quad \text{a.e. in } \Omega. \quad (1.87)$$

Thus

$$\int_\Omega |\nabla(u_1 - u_2)|^2\varphi'^2(u_1 - u_2) + a\varphi(u_1 - u_2)^2 \, dx = 0$$
$$\Rightarrow \quad |\varphi(u_1 - u_2)|_{a,\Omega} = 0 \quad \Rightarrow \quad \varphi(u_1 - u_2) = 0$$

and (1.81) follows. This completes the proof of the theorem. $\qquad \square$

PROOF OF THE LEMMA. Let v_n be a sequence of $C^1(\overline{\Omega})$-functions belonging to V. It is clear that

$$\varphi(v_n) \in V,$$

since it holds, for derivatives taken in the usual or distributional sense:

$$|\varphi(v_n)| \leq c|v_n|, \qquad |\partial_{x_i}\varphi(v_n)| = |\varphi'(v_n)\partial_{x_i}v_n| \leq c|\partial_{x_i}v_n|.$$

Suppose now that $v_n \to v$ in V – i.e., in $H^1(\Omega)$. From (1.82) we derive

$$|\varphi(v_n) - \varphi(v_m)| \leq c|v_n - v_m| \quad \forall m, n$$
$$\Rightarrow \quad |\varphi(v_n) - \varphi(v_m)|_{2,\Omega} \leq c|v_n - v_m|_{2,\Omega} \quad \forall m, n. \quad (1.88)$$

Since $v_n \to v$ in V, $\varphi(v_n)$ is a Cauchy sequence in $L^2(\Omega)$. It follows that $\varphi(v_n)$ converges in $L^2(\Omega)$ toward a function w. Up to a subsequence we can assume that

$$v_n \to v \quad \text{a.e. in } \Omega, \qquad \varphi(v_n) \to w \quad \text{a.e. in } \Omega.$$

From the continuity of φ we then derive that $w = \varphi(v)$ and thus $\varphi(v_n) \to \varphi(v)$ in $L^2(\Omega)$. Next we have

$$\begin{aligned}
|\partial_{x_i}(\varphi(v_n)) - \varphi'(v)\partial_{x_i}v|_{2,\Omega} &= |\varphi'(v_n)\partial_{x_i}v_n - \varphi'(v)\partial_{x_i}v|_{2,\Omega} \\
&\leq |\varphi'(v_n)(\partial_{x_i}v_n - \partial_{x_i}v) + (\varphi'(v_n) - \varphi'(v))\partial_{x_i}v|_{2,\Omega} \qquad (1.89) \\
&\leq c|\partial_{x_i}v_n - \partial_{x_i}v|_{2,\Omega} + |(\varphi'(v_n) - \varphi'(v))\partial_{x_i}v|_{2,\Omega}.
\end{aligned}$$

Suppose that

$$|(\varphi'(v_n) - \varphi'(v))\partial_{x_i}v|_{2,\Omega} \not\to 0$$

when $n \to +\infty$. Then, for every $\varepsilon > 0$ there exists a subsequence – still labeled by n – such that

$$|(\varphi'(v_n) - \varphi'(v))\partial_{x_i}v|_{2,\Omega} \geq \varepsilon. \qquad (1.90)$$

We know that – up to a subsequence

$$v_n \to v \quad \text{a.e. in } \Omega \quad \Rightarrow \quad \varphi'(v_n) \to \varphi'(v) \quad \text{a.e. in } \Omega.$$

Then, from the Lebesgue theorem we deduce

$$\varphi'(v_n)\partial_{x_i}v \to \varphi'(v)\partial_{x_i}v$$

which contradicts (1.90). Thus from (1.89) we deduce that

$$\partial_{x_i}\varphi(v_n) \to \varphi'(v)\partial_{x_i}v \quad \text{in } L^2(\Omega)$$

and also in $\mathcal{D}'(\Omega)$. Since in $L^2(\Omega)$ – and also in $\mathcal{D}'(\Omega)$ –

$$\varphi(v_n) \to \varphi(v)$$

we have

$$\partial_{x_i}\varphi(v_n) \to \partial_{x_i}\varphi(v) \quad \text{in } \mathcal{D}'(\Omega).$$

Thus it holds in the distributional sense that

$$\partial_{x_i}\varphi(v) = \varphi'(v)\partial_{x_i}v.$$

This shows that $\varphi(v) \in H^1(\Omega)$ and $\varphi(v_n) \to \varphi(v)$ in $H^1(\Omega)$. Since $\varphi(v_n) \in V$ we have $\varphi(v) \in V$ and the proof of the lemma is complete. $\qquad\square$

Chapter 2

Some Model Techniques

In this chapter we would like to explain for simple problems the key ideas allowing us to describe the asymptotic behaviour of problems set in domains having a dimension becoming unbounded. We will restrict ourselves to linear elliptic problems.

2.1. The case of lateral Dirichlet boundary conditions on a rectangle

Let $\ell > 0$ be a real number. We denote by Ω_ℓ the rectangle $(-\ell, \ell) \times (-1, 1)$ depicted in Figure 2.1. Let ω be the interval $(-1, 1)$. Let $f = f(x_2)$ be a function

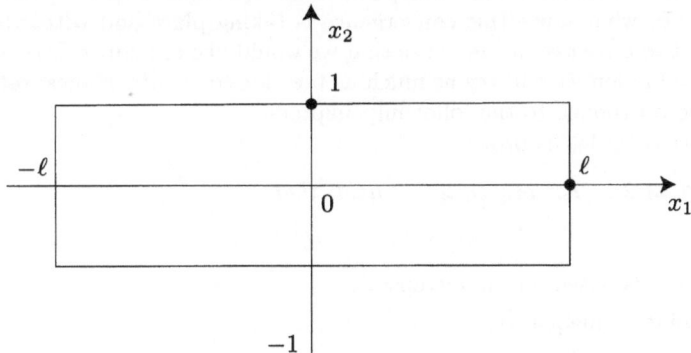

FIGURE 2.1

of x_2 only. For instance suppose $f \in L^2(\omega)$. Then, according to Theorem 1.9 there exists a unique weak solution to the problem

$$\begin{cases} -\Delta u_\ell = f & \text{in } \Omega_\ell, \\ u_\ell = 0 & \text{on } \partial\Omega_\ell. \end{cases} \qquad (2.1)$$

Above Δ is the usual Laplace operator defined by

$$\Delta = \partial^2_{x_1} + \partial^2_{x_2}$$

where $\partial_{x_i}^2$ denotes the second derivative in the direction x_i. Recall that by a weak solution we mean a function u_ℓ solution to

$$\begin{cases} \displaystyle\int_{\Omega_\ell} \nabla u_\ell \cdot \nabla v \, dx = \int_{\Omega_\ell} f(x_2)v(x) \, dx \quad \forall\, v \in H_0^1(\Omega_\ell), \\ u_\ell \in H_0^1(\Omega_\ell). \end{cases} \tag{2.2}$$

Similarly, by Theorem 1.9 there exists a unique u_∞ solution to

$$\begin{cases} \displaystyle\int_\omega \partial_{x_2} u_\infty \partial_{x_2} v \, dx_2 = \int_\omega f v \, dx_2 \quad \forall\, v \in H_0^1(\omega), \\ u_\infty \in H_0^1(\omega), \end{cases} \tag{2.3}$$

i.e., a unique weak solution to

$$\begin{cases} -\partial_{x_2}^2 u_\infty = f & \text{in } \omega, \\ u_\infty = 0 & \text{on } \partial\omega = \{-1, 1\}. \end{cases} \tag{2.4}$$

(For convenience we denote by x_2 the variable in $(-1, 1)$.)

Since f is independent of x_1, it is reasonable to expect that, when $\ell \to +\infty$, u_ℓ converges toward something independent of x_1 and a natural candidate is of course u_∞. In what sense this convergence is taking place and with what speed with respect to ℓ are two of the issues that we would like to address here. Of course in this introduction we will try as much as possible to avoid technical refinements that will be postponed to the following chapters.

To start with, let us prove:

THEOREM 2.1. *For any $\ell_0 > 0$ it holds that*

$$u_\ell \to u_\infty \quad \text{in } H^1(\Omega_{\ell_0}). \tag{2.5}$$

The proof is based on three steps:

a) a Poincaré inequality,

b) an estimate of u_ℓ,

c) the use of a suitable test function.

Let us start with step a) and prove

LEMMA 2.2. *For any $v \in H^1(\Omega_\ell)$ such that $v = 0$ on $(-\ell, \ell) \times \{-1\}$ it holds that*

$$|v|_{2,\Omega_\ell} \leq \sqrt{2}|\partial_{x_2} v|_{2,\Omega_\ell}. \tag{2.6}$$

(Note that the constant $\sqrt{2}$ is independent of ℓ.)

PROOF OF LEMMA 2.2. If $\Gamma_0 = (-\ell, \ell) \times \{-1\}$, by a function of $H^1(\Omega_\ell)$ vanishing on Γ_0 we just mean a function of $H_0^1(\Omega_\ell; \Gamma_0)$ – see (1.36). Let v be a smooth function of $H_0^1(\Omega_\ell; \Gamma_0)$. We have for any $(x_1, x_2) \in \Omega_\ell$

$$v(x_1, x_2) = \int_{-1}^{x_2} \partial_{x_2} v(x_1, s) \, ds. \tag{2.7}$$

Taking absolute values and using the Cauchy–Schwarz inequality,

$$|v(x_1, x_2)| \leq \int_{-1}^{x_2} |\partial_{x_2} v(x_1, s)| \, ds$$

$$\leq \left\{ \int_{-1}^{x_2} |\partial_{x_2} v(x_1, s)|^2 \, ds \right\}^{1/2} (x_2 + 1)^{1/2} \qquad (2.8)$$

$$\leq \left\{ \int_{-1}^{1} |\partial_{x_2} v(x_1, s)|^2 \, ds \right\}^{1/2} (x_2 + 1)^{1/2}.$$

Squaring both sides of the inequality and integrating in x_2 we obtain

$$\int_{-1}^{1} v(x_1, x_2)^2 \, dx_2 \leq \left\{ \int_{-1}^{1} \partial_{x_2} v(x_1, x_2)^2 \, dx_2 \right\} \frac{(x_2 + 1)^2}{2} \Big|_{-1}^{1}$$

$$= 2 \int_{-1}^{1} \partial_{x_2} v(x_1, x_2)^2 \, dx_2.$$

Integrating between $-\ell, \ell$ and taking the square root we get

$$|v|_{2,\Omega_\ell} \leq \sqrt{2} |\partial_{x_2} v|_{2,\Omega_\ell}.$$

The result follows then by density of the smooth functions in $H_0^1(\Omega_\ell; \Gamma_0)$ – see (1.28). $\qquad \square$

We can then pass to point b) and get an estimate for u_ℓ.

LEMMA 2.3. *Let u_ℓ, u_∞ be the functions introduced above. It holds that*

$$||\nabla u_\ell||_{2,\Omega_\ell} \leq \sqrt{2\ell} |\partial_{x_2} u_\infty|_{2,\omega}. \qquad (2.9)$$

PROOF. For any $v \in \mathcal{D}(\Omega_\ell)$ by (2.2), (2.3) we have (note that at this stage we are trying to avoid the use of Proposition 3.1 below)

$$\int_{\Omega_\ell} \nabla u_\ell \nabla v \, dx = \int_{(-\ell,\ell)} \int_\omega f(x_2) v(x_1, x_2) \, dx_2 \, dx_1,$$

$$= \int_{(-\ell,\ell)} \int_\omega \partial_{x_2} u_\infty(x_2) \partial_{x_2} v(x_1, x_2) \, dx_2 \, dx_1. \qquad (2.10)$$

By density of $\mathcal{D}(\Omega_\ell)$ in $H_0^1(\Omega_\ell)$, this equality holds for $v \in H_0^1(\Omega_\ell)$ and in particular for $v = u_\ell$ and we get

$$||\nabla u_\ell||_{2,\Omega_\ell}^2 = \int_{\Omega_\ell} \partial_{x_2} u_\infty \partial_{x_2} u_\ell \, dx \leq |\partial_{x_2} u_\infty|_{2,\Omega_\ell} |\partial_{x_2} u_\ell|_{2,\Omega_\ell} \qquad (2.11)$$

by the Cauchy–Schwarz inequality. Since

$$|\partial_{x_2} u_\ell|_{2,\Omega_\ell} \leq ||\nabla u_\ell||_{2,\Omega_\ell}$$

we derive

$$||\nabla u_\ell||_{2,\Omega_\ell} \leq |\partial_{x_2} u_\infty|_{2,\Omega_\ell} = \left\{ \int_{(-\ell,\ell)} \int_\omega \partial_{x_2} u_\infty^2(x_2) \, dx_2 \, dx_1 \right\}^{1/2} = \sqrt{2\ell} |\partial_{x_2} u_\infty|_{2,\omega}$$

which completes the proof of the lemma. $\qquad \square$

We can then conclude:

PROOF OF THEOREM 2.1. Let us introduce ϱ the function depicted in Figure 2.2. Clearly

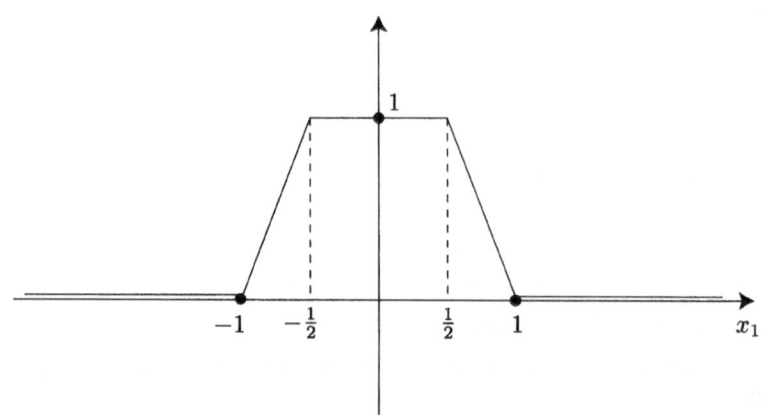

FIGURE 2.2

$$(u_\ell - u_\infty)\varrho^2\left(\frac{x_1}{\ell}\right) \in H_0^1(\Omega_\ell) \tag{2.12}$$

so that by (2.2), (2.3) it holds that (see below (2.10)):

$$\int_{\Omega_\ell} \nabla u_\ell \nabla \left\{ (u_\ell - u_\infty)\varrho^2\left(\frac{x_1}{\ell}\right) \right\} dx$$
$$= \int_{\Omega_\ell} \partial_{x_2} u_\infty \partial_{x_2}(u_\ell - u_\infty)\varrho^2\left(\frac{x_1}{\ell}\right) dx. \tag{2.13}$$

(Note that $\varrho(\frac{x_1}{\ell})$ is independent of x_2.) Performing the derivatives in the first integral we obtain

$$\int_{\Omega_\ell} \partial_{x_1} u_\ell \partial_{x_1}(u_\ell - u_\infty)\varrho^2\left(\frac{x_1}{\ell}\right) dx + \int_{\Omega_\ell} \partial_{x_2} u_\ell \partial_{x_2}(u_\ell - u_\infty)\varrho^2\left(\frac{x_1}{\ell}\right) dx$$
$$= \int_{\Omega_\ell} \partial_{x_2} u_\infty \partial_{x_2}(u_\ell - u_\infty)\varrho^2\left(\frac{x_1}{\ell}\right) dx$$
$$- \frac{2}{\ell} \int_{\Omega_\ell} \partial_{x_1} u_\ell \partial_{x_1}\varrho\left(\frac{x_1}{\ell}\right)(u_\ell - u_\infty)\varrho\left(\frac{x_1}{\ell}\right) dx. \tag{2.14}$$

Since u_∞ is independent of x_1 this can also be written

$$\int_{\Omega_\ell} |\nabla(u_\ell - u_\infty)|^2 \varrho^2\left(\frac{x_1}{\ell}\right) dx$$

$$= -\frac{2}{\ell}\int_{\Omega_\ell} \partial_{x_1} u_\ell \partial_{x_1}\varrho\left(\frac{x_1}{\ell}\right)(u_\ell - u_\infty)\varrho\left(\frac{x_1}{\ell}\right) dx. \tag{2.15}$$

It is clear that $|\partial_{x_1}\varrho| \le 2$. So, by the Cauchy–Schwarz inequality, we derive from (2.15)

$$\int_{\Omega_\ell} |\nabla(u_\ell - u_\infty)|^2 \varrho^2\left(\frac{x_1}{\ell}\right) dx \le \frac{4}{\ell}|\partial_{x_1} u_\ell|_{2,\Omega_\ell}\left|(u_\ell - u_\infty)\varrho\left(\frac{x_1}{\ell}\right)\right|_{2,\Omega_\ell}. \tag{2.16}$$

Applying then Lemma 2.2 for $v = (u_\ell - u_\infty)\varrho$ we deduce

$$\left\|\nabla(u_\ell - u_\infty)|\varrho\left(\frac{x_1}{\ell}\right)\right|^2_{2,\Omega_\ell} \le \frac{4\sqrt{2}}{\ell}|\partial_{x_1} u_\ell|_{2,\Omega_\ell}\left|\partial_{x_2}(u_\ell - u_\infty)\varrho\left(\frac{x_1}{\ell}\right)\right|_{2,\Omega_\ell}$$

$$\le \frac{4\sqrt{2}}{\ell}|\partial_{x_1} u_\ell|_{2,\Omega_\ell}\left\||\nabla(u_\ell - u_\infty)|\varrho\left(\frac{x_1}{\ell}\right)\right\|_{2,\Omega_\ell}. \tag{2.17}$$

From this it follows that

$$\left\||\nabla(u_\ell - u_\infty)|\varrho\left(\frac{x_1}{\ell}\right)\right\|_{2,\Omega_\ell} \le \frac{4\sqrt{2}}{\ell}|\partial_{x_1} u_\ell|_{2,\Omega_\ell} \le \frac{8}{\sqrt{\ell}}|\partial_{x_2} u_\infty|_{2,\omega} \tag{2.18}$$

by Lemma 2.3. Since $\varrho(\frac{x_1}{\ell}) = 1$ on $\Omega_{\ell/2}$ we have obtained

$$\left\{\int_{\Omega_{\ell/2}} |\nabla(u_\ell - u_\infty)|^2 dx\right\}^{\frac{1}{2}} \le \frac{8}{\sqrt{\ell}}|\partial_{x_2} u_\infty|_{2,\omega}$$

and for $\frac{\ell}{2} > \ell_0$

$$\|\nabla(u_\ell - u_\infty)\|_{2,\Omega_{\ell_0}} \le \frac{8}{\sqrt{\ell}}|\partial_{x_2} u_\infty|_{2,\omega}. \tag{2.19}$$

The result is then a trivial consequence of Theorem 1.4. $\square$

REMARK 2.1. The fact is that $f \in L^2(\omega)$ did not play any role since we went around f by the use of u_∞ (see (2.10)). We will come back to this in the following.

2.2. The case of lateral Neumann boundary conditions on a rectangle

Consider Ω_ℓ, f as in the preceding section. Somehow, our proof there was based on an estimate of the $L^2(\Omega_\ell)$-norm in terms of the norm defined by the operator in ω (Cf. (2.6)). This estimate will hold also in the case that we would like to address now. Thus, consider u_ℓ the weak solution to

$$\begin{cases} -\Delta u_\ell + a u_\ell = f \text{ in } \Omega_\ell, \\ u_\ell = 0 \text{ on } \{-\ell, \ell\} \times \omega, \quad \dfrac{\partial u_\ell}{\partial \nu} = 0 \text{ on } (-\ell, \ell) \times \partial\omega, \end{cases} \tag{2.20}$$

where $a = a(x_2)$ belongs to $L^\infty(\omega)$. In other words setting

$$\Gamma_0 = \{-\ell, \ell\} \times \omega \tag{2.21}$$

for $V = H_0^1(\Omega_\ell; \Gamma_0)$, $f \in V'$ – see Theorem 1.10 – u_ℓ is the solution to

$$\begin{cases} \displaystyle\int_{\Omega_\ell} \nabla u_\ell \cdot \nabla v + a u_\ell v \, dx = \langle f, v \rangle \quad \forall\, v \in V, \\ v \in V. \end{cases} \tag{2.22}$$

2.2.1. The case where $a \not\equiv 0$. Before we continue, let us analyse how $f = f(x_2)$ allows us to define an element of V'. In fact let us do it directly for an element

$$f \in (H^1(\omega))', \tag{2.23}$$

where $H^1(\omega)'$ denotes the dual of $H^1(\omega)$. Let v be a smooth function of $C_0^1(\overline{\Omega}_\ell; \Gamma_0)$ – see (1.28). Then, for every $x_1 \in (-\ell, \ell)$

$$\langle f, v(x_1, \cdot) \rangle \tag{2.24}$$

is well defined. Moreover, it is easy to show that this function is measurable in x_1 – it is even continuous and it holds that

$$|\langle f, v(x_1, \cdot) \rangle| \le |f|_* |v(x_1, \cdot)|_{1,2}. \tag{2.25}$$

($|\cdot|_{1,2}$ denotes here the $H^1(\omega)$-norm, $|\cdot|_*$ the strong dual norm in $H^1(\omega)'$.) It follows that the function (2.24) belongs to $L^2(-\ell, \ell) \subset L^1(-\ell, \ell)$. Thus, we can define $\langle f, v \rangle$ in the duality V', V by setting

$$\langle f, v \rangle = \int_{(-\ell, \ell)} \langle f, v(x_1, \cdot) \rangle \, dx_1. \tag{2.26}$$

We have indeed

$$\begin{aligned} |\langle f, v \rangle| &\le \int_{(-\ell, \ell)} |\langle f, v(x_1, \cdot) \rangle| \, dx_1 \\ &\le \int_{(-\ell, \ell)} |f|_* |v(x_1, \cdot)|_{1,2} \, dx_1 \\ &\le |f|_* \int_{(-\ell, \ell)} |v(x_1, \cdot)|_{1,2} \, dx_1 \\ &\le |f|_* \sqrt{2\ell} \left\{ \int_{(-\ell, \ell)} |v(x_1, \cdot)|_{1,2}^2 \, dx_1 \right\}^{\frac{1}{2}} \quad \text{(Cauchy–Schwarz)} \\ &\le \sqrt{2\ell} |f|_* \left\{ \int_{(-\ell, \ell)} \int_\omega \partial_{x_2} v(x_1, x_2)^2 + v(x_1, x_2)^2 \, dx_1 \, dx_2 \right\}^{\frac{1}{2}}. \end{aligned} \tag{2.27}$$

So, the linear form (2.26) defined on smooth functions is continuous on V and thus admits a unique extension to V defined also by (2.26). We have thus defined u_ℓ

the solution to (2.22) for $f \in (H^1(\omega))'$. Since $a = a(x_2) \not\equiv 0$, it is clear that there exists a unique u_∞ solution to

$$\begin{cases} \int_\omega \partial_{x_2} u_\infty \partial_{x_2} v + a u_\infty v \, dx = \langle f, v \rangle \quad \forall v \in H^1(\omega), \\ u_\infty \in H^1(\omega). \end{cases} \tag{2.28}$$

We would like to show that u_ℓ converges toward u_∞ when $\ell \to +\infty$. Indeed we have

THEOREM 2.4. *For any $\ell_0 > 0$, if u_ℓ, u_∞ are the solutions to (2.22), (2.28), it holds that*

$$u_\ell \to u_\infty \quad \text{in } H^1(\Omega_{\ell_0}). \tag{2.29}$$

We follow the same steps as in the preceding section. First we have:

LEMMA 2.5. *For every $v \in H^1(\Omega_\ell)$ it holds that*

$$|v|_{2,\Omega_\ell} \leq C \left\{ \int_{\Omega_\ell} (\partial_{x_2} v)^2 + a v^2 \, dx \right\}^{\frac{1}{2}} \tag{2.30}$$

where C is a constant independent of ℓ and ∂_{x_2} denotes as before the derivative in the direction x_2.

PROOF. Let v be a smooth function. From Theorem 1.3 applied for $\Omega = \omega$ we derive

$$|v(x_1, \cdot)|^2_{2,\omega} \leq C \int_\omega (\partial_{x_2} v(x_1, x_2))^2 + a(x_2) v^2 (x_1, x_2) \, dx_2.$$

Integrating in x_1 on $(-\ell, \ell)$ we get

$$|v|^2_{2,\Omega_\ell} \leq C \int_{\Omega_\ell} (\partial_{x_2} v)^2 + a v^2 \, dx. \tag{2.31}$$

The result follows by density of the smooth functions in $H^1(\Omega_\ell)$. $\square$

Next we estimate u_ℓ. We have

LEMMA 2.6. *Let u_ℓ, u_∞ be the solutions to (2.22), (2.28). It holds that*

$$|u_\ell|_{a,\Omega_\ell} \leq \sqrt{2\ell} |u_\infty|_{a,\omega}. \tag{2.32}$$

($|u_\ell|_{a,K}$ is the norm defined by (1.19) when $\Omega = K$.)

PROOF. Let us take $v = u_\ell$ in (2.22). Taking into account (2.26), (2.28) we obtain

$$|u_\ell|^2_{a,\Omega_\ell} = \int_{(-\ell,\ell)} \int_\omega \partial_{x_2} u_\infty \partial_{x_2} u_\ell + a u_\infty u_\ell \, dx \leq \int_{(-\ell,\ell)} |u_\infty|_{a,\omega} |u_\ell(x_1, \cdot)|_{a,\omega} \, dx_1.$$

Applying the Cauchy–Schwarz inequality this leads to

$$|u_\ell|^2_{a,\Omega_\ell} \leq |u_\infty|_{a,\omega} \left\{ \int_{(-\ell,\ell)} |u_\ell(x_1, \cdot)|^2_{a,\omega} \, dx_1 \right\}^{\frac{1}{2}} \sqrt{2\ell} \leq \sqrt{2\ell} |u_\infty|_{a,\omega} |u_\ell|_{a,\Omega_\ell} \tag{2.33}$$

and the result follows. $\square$

We can then conclude

PROOF OF THEOREM 2.4. If ϱ is the function defined on Figure 2.2 we have now

$$(u_\ell - u_\infty)\varrho^2\left(\frac{x_1}{\ell}\right) \in V. \tag{2.34}$$

Thus from (2.22), (2.26), (2.28) we derive

$$\int_{\Omega_\ell} \nabla u_\ell \nabla\left\{(u_\ell - u_\infty)\varrho^2\left(\frac{x_1}{\ell}\right)\right\} + au_\ell(u_\ell - u_\infty)\varrho^2\left(\frac{x_1}{\ell}\right) dx$$

$$= \int_{(-\ell,\ell)} \int_\omega \partial_{x_2} u_\infty \partial_{x_2}(u_\ell - u_\infty)\varrho^2\left(\frac{x_1}{\ell}\right) + au_\infty(u_\ell - u_\infty)\varrho^2\left(\frac{x_1}{\ell}\right) dx. \tag{2.35}$$

Proceeding as in (2.14), (2.15) we obtain easily

$$\int_{\Omega_\ell} \{|\nabla(u_\ell - u_\infty)|^2 + a(u_\ell - u_\infty)^2\}\varrho^2\left(\frac{x_1}{\ell}\right) dx$$

$$= -\frac{2}{\ell}\int_{\Omega_\ell} \partial_{x_1} u_\ell \partial_{x_1}\varrho\left(\frac{x_1}{\ell}\right)(u_\ell - u_\infty)\varrho\left(\frac{x_1}{\ell}\right) dx \tag{2.36}$$

$$\leq \frac{4}{\ell}|\partial_{x_1} u_\ell|_{2,\Omega_\ell}\left|(u_\ell - u_\infty)\varrho\left(\frac{x_1}{\ell}\right)\right|_{2,\Omega_\ell}.$$

Applying Lemma 2.5 we obtain for some constant C independent of ℓ

$$\int_{\Omega_\ell} \{|\nabla(u_\ell - u_\infty)|^2 + a(u_\ell - u_\infty)^2\}\varrho^2\left(\frac{x_1}{\ell}\right) dx$$

$$\leq \frac{C}{\ell}|\partial_{x_1} u_\ell|_{2,\Omega_\ell}\left\{\int_{\Omega_\ell} [\partial_{x_2}(u_\ell - u_\infty)^2 + a(u_\ell - u_\infty)^2]\varrho^2\left(\frac{x_1}{\ell}\right) dx\right\}^{\frac{1}{2}}$$

from which we derive

$$\left(\int_{\Omega_\ell} \{|\nabla(u_\ell - u_\infty)|^2 + a(u_\ell - u_\infty)^2\}\varrho^2\left(\frac{x_1}{\ell}\right) dx\right)^{\frac{1}{2}}$$

$$\leq \frac{C}{\ell}|\partial_{x_1} u_\ell|_{2,\Omega_\ell} \leq \frac{C}{\ell}|u_\ell|_{a,\Omega_\ell} \leq \frac{C}{\sqrt{\ell}}|u_\infty|_{a,\omega}. \tag{2.37}$$

(C denotes different constants independent of ℓ). Taking $\frac{\ell}{2} > \ell_0$ we derive

$$|u_\ell - u_\infty|_{a,\Omega_{\ell_0}} \leq \frac{C}{\sqrt{\ell}}|u_\infty|_{a,\omega}. \tag{2.38}$$

The result follows then from Theorem 1.3.   □

2.2.2. The case where $a \equiv 0$. In this case consider first that u_ℓ, f, a are smooth enough so that u_ℓ is a classical solution to

$$\begin{cases} -\partial_{x_1}^2 u_\ell - \partial_{x_2}^2 u_\ell = f & \text{in } \Omega_\ell, \\ u_\ell = 0 \text{ on } \{-\ell, \ell\} \times \omega, \quad \partial_{x_2} u_\ell = 0 \text{ on } (-\ell, \ell) \times \partial\omega. \end{cases} \tag{2.39}$$

(We refer the reader to [**19**] for regularity results on these problems.) Assuming this is possible, let us integrate the first equation of (2.39) over $\omega = (-1, 1)$. We get

$$-\partial_{x_1}^2 \int_\omega u_\ell(x_1, x_2) \, dx_2 = \int_\omega f \, dx_2 \quad \text{for } x_1 \in (-\ell, \ell). \tag{2.40}$$

It follows that for some constants α, β,

$$\int_\omega u_\ell(x_1, x_2) \, dx_2 = -\frac{1}{2} x_1^2 \int_\omega f \, dx_2 + \alpha x_1 + \beta. \tag{2.41}$$

Since $u_\ell = 0$ on $\{-\ell, \ell\} \times \omega$, letting $x_1 = \pm\ell$ in the above equality we derive

$$\alpha = 0, \qquad \beta = \frac{1}{2} \ell^2 \int_\omega f \, dx_2. \tag{2.42}$$

We obtain then

$$\int_\omega u_\ell(x_1, x_2) \, dx_2 = \frac{1}{2}(\ell^2 - x_1^2) \int_\omega f \, dx_2. \tag{2.43}$$

If

$$\int_\omega f \, dx_2 \neq 0 \tag{2.44}$$

the above quantity converges toward $\pm\infty$ when $\ell \to +\infty$ and there is then no hope to get convergence of u_ℓ. Note also that in the case of (2.44) the solution u_∞ does not exist – see (1.72). Thus we consider now $f \in (H^1(\omega))'$ with

$$\langle f, 1 \rangle = 0 \tag{2.45}$$

and we introduce u_∞ solution to

$$\begin{cases} \int_\omega \partial_{x_2} u_\infty \partial_{x_2} v \, dx = \langle f, v \rangle \quad \forall v \in H^1(\omega), \\ u_\infty \in H^1(\omega), \quad \int_\omega u_\infty \, dx_2 = 0, \end{cases} \tag{2.46}$$

(see Theorem 1.13). Then we can show

THEOREM 2.7. *Let u_ℓ, u_∞ be the solutions to* (2.22), (2.46) *under the assumptions $a = 0$,* (2.45). *For any $\ell_0 > 0$ it holds that*

$$u_\ell \to u_\infty \quad \text{in } H^1(\Omega_{\ell_0}). \tag{2.47}$$

We start with the following lemma:

LEMMA 2.8. *Let $v \in H^1(\Omega_\ell)$ such that*

$$\int_\omega v(x_1, x_2) \, dx_2 = 0 \quad a.e. \ x_1 \in (-\ell, \ell); \tag{2.48}$$

then there exists a constant C independent of ℓ such that

$$|v|_{2,\Omega_\ell} \le C \|\partial_{x_2} v\|_{2,\Omega_\ell}. \tag{2.49}$$

PROOF. If $v \in H^1(\Omega_\ell)$, then for a.e. x_1

$$v(x_1, \cdot) \in H^1(\omega). \tag{2.50}$$

Indeed, $v \in H^1(\Omega_\ell)$ implies that

$$\int_{\Omega_\ell} \partial_{x_2} v^2 + v^2 \, dx < +\infty$$

hence by Fubini's Theorem

$$\int_\omega \partial_{x_2} v^2(x_1, x_2) + v^2(x_1, x_2) \, dx_2 < +\infty$$

for a.e. x_1. Thus, for a.e. $x_1 \in (-\ell, \ell)$, it holds that

$$\partial_{x_2} v(x_1, \cdot), v(x_1, \cdot) \in L^2(\omega).$$

It remains to show that the above derivative is the derivative of $v(x_1, \cdot)$ in the distributional sense in ω. For that – and by definition of $\partial_{x_2} v$ – we notice that

$$\int_{\Omega_\ell} \partial_{x_2} v \psi \, dx = -\int_{\Omega_\ell} v \partial_{x_2} \psi \, dx \quad \forall \psi \in \mathcal{D}(\Omega_\ell).$$

Choosing $\psi(x_1, x_2) = \varrho(x_1) \varphi(x_2)$ with $\varrho \in \mathcal{D}(-\ell, \ell)$, $\varphi \in \mathcal{D}(\omega)$ we derive

$$\int_{(-\ell, \ell)} \varrho(x_1) \int_\omega \partial_{x_2} v(x_1, x_2) \varphi(x_2) \, dx_2 \, dx_1$$

$$= -\int_{(-\ell, \ell)} \varrho(x_1) \int_\omega v(x_1, x_2) \partial_{x_2} \varphi(x_2) \, dx_2 \, dx_1.$$

Hence for a.e. $x_1 \in (-\ell, \ell)$

$$\int_\omega \partial_{x_2} v(x_1, x_2) \varphi(x_2) \, dx_2 = -\int_\omega v(x_1, x_2) \partial_{x_2} \varphi(x_2) \, dx_2$$

which completes the proof of (2.50). Next, applying Theorem 1.7 we obtain for a.e. x_1

$$|v(x_1, \cdot)|_{2,\omega} \le C |\partial_{x_2} v(x_1, \cdot)|_{2,\omega}$$

where C is some constant depending on ω only. Squaring this inequality and integrating in x_1 on $(-\ell, \ell)$ we obtain (2.49). This completes the proof of the lemma. $\qquad \square$

Next, we can estimate u_ℓ. First we remark that:

LEMMA 2.9. *For $f \in H^1(\omega)'$ let u_ℓ be the solution to*

$$\begin{cases} \iint_{\Omega_\ell} \nabla u_\ell \nabla v \, dx = \langle f, v \rangle \; \forall v \in H^1(\Omega_\ell), \quad v = 0 \text{ on } \{-\ell, \ell\} \times \omega, \\ u_\ell \in H^1(\Omega_\ell), \quad u_\ell = 0 \text{ on } \{-\ell, \ell\} \times \omega, \end{cases} \tag{2.51}$$

(recall (2.26)). Then, for f satisfying (2.45), it holds that

$$\int_\omega u_\ell(x_1, x_2) \, dx_2 = 0 \quad a.e. \; x_1 \in (-\ell, \ell). \tag{2.52}$$

PROOF. Consider $\varrho = \varrho(x_1) \in \mathcal{D}(-\ell, \ell)$. Plugging this function into (2.51) we obtain by (2.26)

$$\int_{\Omega_\ell} \partial_{x_1} u_\ell \partial_{x_1} \varrho \, dx = \int_{(-\ell,\ell)} \varrho(x_1) \langle f, 1 \rangle \, dx_1 = 0.$$

It follows that a.e. x_1

$$\int_\omega \partial_{x_1} u_\ell(x_1, x_2) \, dx_2 = \text{cst} = c_1.$$

Since a.e. x_2, $u(\cdot, x_2) \in H^1(-\ell, \ell)$ – see above – we derive that

$$\partial_{x_1} \int_\omega u_\ell(x_1, x_2) \, dx_2 = c_1$$

$$\Rightarrow \qquad \int_\omega u_\ell(x_1, x_2) \, dx_2 = c_1 x_1 + c_2$$

where c_1, c_2 are some constants. Since u_ℓ vanishes on $\{-\ell, \ell\} \times \omega$ we derive easily that $c_1 = c_2 = 0$ and (2.52) follows. $\qquad \square$

We can then prove

LEMMA 2.10. *Let u_ℓ be the solution to (2.51) with f satisfying (2.45). Then, it holds that*

$$\|\nabla u_\ell\|_{2,\Omega_\ell} \le \sqrt{2\ell} |f|_* \tag{2.53}$$

where $| \cdot |_*$ denotes the strong dual norm in W' for $W = W(\omega; \omega)$ equipped with the norm $|\partial_{x_2} v|_{2,\omega}$ (see (1.38)).

PROOF. We take $v = u_\ell$ in (2.51). It follows that

$$\|\nabla u_\ell\|_{2,\Omega_\ell}^2 = \int_{(-\ell,\ell)} \langle f, u_\ell(x_1, \cdot) \rangle \, dx_1$$

$$\le \int_{(-\ell,\ell)} |f|_* |\partial_{x_2} u_\ell(x_1, \cdot)|_{2,\omega} \, dx_1$$

(recall Theorem 1.7).

Using the Cauchy–Schwarz inequality it follows that

$$\|\nabla u_\ell\|_{2,\Omega_\ell}^2 \le |f|_* |\partial_{x_2} u_\ell|_{2,\Omega_\ell} \sqrt{2\ell} \le \sqrt{2\ell} |f|_* \|\nabla u_\ell\|_{2,\Omega_\ell} \tag{2.54}$$

which completes the proof. $\qquad \square$

We can then complete the proof of Theorem 2.7.

PROOF OF THEOREM 2.7. Let ϱ be the function introduced in Figure 2.2. We have

$$v = (u_\ell - u_\infty)\varrho^2\left(\frac{x_1}{\ell}\right) \in H^1(\Omega_\ell). \tag{2.55}$$

Moreover, by (2.46), (2.52) it holds that

$$\int_\omega v(x_1, x_2)\, dx_2 = 0 \quad \text{a.e. } x_1 \in (-\ell, \ell).$$

Thus, combining (2.51), (2.26), (2.46) we obtain

$$\int_{\Omega_\ell} \nabla u_\ell \nabla v\, dx = \int_{(-\ell,\ell)} \int_\omega \partial_{x_2} u_\infty \partial_{x_2} v\, dx$$

$$\Leftrightarrow \quad \int_{\Omega_\ell} \nabla u_\ell \nabla\left\{(u_\ell - u_\infty)\varrho^2\left(\frac{x_1}{\ell}\right)\right\} dx = \int_{\Omega_\ell} \partial_{x_2} u_\infty \partial_{x_2}(u_\ell - u_\infty)\varrho^2\left(\frac{x_1}{\ell}\right) dx.$$

Proceeding as usual – i.e., expanding the derivations in the first integral, we obtain since u_∞ is independent of x_1

$$\int_{\Omega_\ell} |\nabla(u_\ell - u_\infty)|^2\varrho^2\left(\frac{x_1}{\ell}\right) dx = -\frac{2}{\ell}\int_{\Omega_\ell} \partial_{x_1} u_\ell \partial_{x_1}\varrho\left(\frac{x_1}{\ell}\right)(u_\ell - u_\infty)\varrho\left(\frac{x_1}{\ell}\right) dx$$

$$\leq \frac{4}{\ell}|\partial_{x_1} u_\ell|_{2,\Omega_\ell}\left|(u_\ell - u_\infty)\varrho\left(\frac{x_1}{\ell}\right)\right|_{2,\Omega_\ell} \tag{2.56}$$

by the Cauchy–Schwarz inequality. Using now Lemma 2.8 we get

$$\left||\nabla(u_\ell - u_\infty)|\varrho\left(\frac{x_1}{\ell}\right)\right|^2_{2,\Omega_\ell} \leq \frac{C}{\ell}|\partial_{x_1} u_\ell|_{2,\Omega_\ell}\left|\partial_{x_2}(u_\ell - u_\infty)\varrho\left(\frac{x_1}{\ell}\right)\right|_{2,\Omega_\ell}$$

from which it follows that

$$\left||\nabla(u_\ell - u_\infty)|\varrho\left(\frac{x_1}{\ell}\right)\right|_{2,\Omega_\ell} \leq \frac{C}{\ell}|\partial_{x_1} u_\ell|_{2,\Omega_\ell} \leq \frac{C'}{\sqrt{\ell}}|f|_* \tag{2.57}$$

by Lemma 2.10. Since $\varrho(\frac{x_1}{\ell}) = 1$ on $\Omega_{\ell/2}$ we derive for $\ell/2 > \ell_0$

$$\left||\nabla(u_\ell - u_\infty)|\right|_{2,\Omega_{\ell_0}} \leq \frac{C'}{\sqrt{\ell}}|f|_*.$$

The result follows then from Theorem 1.7 since

$$\int_{\Omega_{\ell_0}} (u_\ell - u_\infty)\, dx = 0.$$

This completes the proof of the theorem. $\square$

REMARK 2.2. In the above section we restricted our analysis to convergence in $H^1(\Omega_{\ell_0})$. Some convergence in other spaces is of course available. Also we will see later on that the rate of convergence in ℓ can be considerably improved. We had here just in mind to convince the reader of the variety of issues that could be attacked with relatively elementary techniques.

2.3. The case of lateral Dirichlet boundary conditions revisited

First let us analyse if a rectangular domain is really necessary for the convergence of u_ℓ toward u_∞. Let us again denote by Ω_ℓ the rectangle $(-\ell, \ell) \times \omega$ where $\omega = (-1, 1)$ and denote by Ω'_ℓ the domain depicted in Figure 2.3. We will suppose

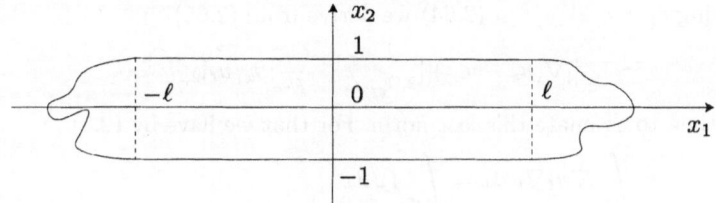

FIGURE 2.3

that
$$\Omega_\ell \subset \Omega'_\ell \subset \Omega_{\ell'} \tag{2.58}$$
for some $\ell' > \ell$. Then, for $f = f(x_2) \in L^2(\omega)$ denote by u_ℓ the solution to
$$\begin{cases} \displaystyle\int_{\Omega'_\ell} \nabla u_\ell \nabla v \, dx = \int_{\Omega'_\ell} fv \, dx, \\ u_\ell \in H_0^1(\Omega'_\ell). \end{cases} \tag{2.59}$$

Then, we have

THEOREM 2.11. *Let us assume that for some $\alpha > 0$*
$$\ell' = O(\ell^\alpha). \tag{2.60}$$
Then under the above assumptions for any $\ell_0 > 0$ it holds that
$$u_\ell \to u_\infty \quad in \ H^1(\Omega_{\ell_0}) \tag{2.61}$$
where u_∞ is the solution to (2.3).

PROOF. Let ϱ be the function introduced in Figure 2.2. Clearly for any $\ell_1 \le \ell$
$$(u_\ell - u_\infty)\varrho^2\left(\frac{x_1}{\ell_1}\right) \in H_0^1(\Omega'_\ell). \tag{2.62}$$

Thus arguing as in (2.13)–(2.17) we obtain
$$\big\||\nabla(u_\ell - u_\infty)|\varrho\big\|_{2,\Omega_{\ell_1}} \le 4\frac{\sqrt{2}}{\ell_1}|\partial_{x_1} u_\ell|_{2,\Omega_{\ell_1}}. \tag{2.63}$$

In particular since $\varrho = 1$ on $\Omega_{\ell_1/2}$

$$||\nabla(u_\ell - u_\infty)||_{2,\Omega_{\ell_1/2}} \le 4\frac{\sqrt{2}}{\ell_1}|\partial_{x_1}u_\ell|_{2,\Omega_{\ell_1}}. \tag{2.64}$$

This implies that for any $\ell_1 \le \ell$ it holds that (recall that u_∞ is independent of x_1)

$$|\partial_{x_1}u_\ell|_{2,\Omega_{\ell_1/2}} \le \frac{4\sqrt{2}}{\ell_1}|\partial_{x_1}u_\ell|_{2,\Omega_{\ell_1}}.$$

Iterating this formula we obtain easily that for any k

$$|\partial_{x_1}u_\ell|_{2,\Omega_{\ell_1/2^k}} \le \frac{C_k}{\ell_1^k}|\partial_{x_1}u_\ell|_{2,\Omega_{\ell_1}}. \tag{2.65}$$

Thus, taking $\ell_1 = \ell/2^{(k-1)}$ in (2.64) we derive from (2.65)

$$||\nabla(u_\ell - u_\infty)||_{2,\Omega_{\ell/2^k}} \le \frac{C_k}{\ell^k}|\partial_{x_1}u_\ell|_{2,\Omega_\ell}. \tag{2.66}$$

We have then to estimate this last norm. For that we have by (2.3)

$$\begin{aligned}\int_{\Omega'_\ell} \nabla u_\ell \nabla v\, dx &= \int_{\Omega'_\ell} fv\, dx \\ &= \int_{\Omega_{\ell'}} \nabla u_\infty \nabla v\, dx \quad \forall v \in H^1_0(\Omega'_\ell)\end{aligned} \tag{2.67}$$

if we suppose v extended by 0 outside Ω'_ℓ to derive these formulae. Taking $v = u_\ell$ we obtain

$$\begin{aligned}||\nabla u_\ell||^2_{2,\Omega'_\ell} &= \int_{\Omega_{\ell'}} \nabla u_\infty \nabla u_\ell\, dx \\ &\le ||\nabla u_\infty||_{2,\Omega_{\ell'}}||\nabla u_\ell||_{2,\Omega_{\ell'}} \\ &= \sqrt{2\ell'}|\partial_{x_2}u_\infty|_{2,\omega}||\nabla u_\ell||_{2,\Omega'_\ell}.\end{aligned} \tag{2.68}$$

From which it follows that

$$||\nabla u_\ell||_{2,\Omega_\ell} \le ||\nabla u_\ell||_{2,\Omega'_\ell} \le \sqrt{2\ell'}|\partial_{x_2}u_\infty|_{2,\omega}. \tag{2.69}$$

Going back to (2.66) and using (2.60) we obtain

$$||\nabla(u_\ell - u_\infty)||_{2,\Omega_{\ell/2^k}} \le \frac{C}{\ell^{k-\frac{\alpha}{2}}}|\partial_{x_2}u_\infty|_{2,\omega} \tag{2.70}$$

for some constant C independent of ℓ. Choosing $k - \frac{\alpha}{2} > 0$, $\frac{\ell}{2^k} > \ell_0$ the result follows. $\square$

REMARK 2.3. We have obtained that for any $r > 0$ there exists a constant C independent of ℓ such that

$$||\nabla(u_\ell - u_\infty)||_{2,\Omega_{\ell_0}} \le \frac{C}{\ell^r} \tag{2.71}$$

i.e., $u_\ell \to u_\infty$ in $H^1(\Omega_{\ell_0})$ with a speed faster than any power of $\frac{1}{\ell}$.

Another question is to know how important are the homogeneous boundary conditions at the ends of the rectangle Ω_ℓ. To see that this does not play a too important rôle consider g a function of x_2 only such that

$$g \in H_0^1(\omega). \tag{2.72}$$

Let then u_ℓ be the solution to

$$\begin{cases} -\Delta u_\ell = f & \text{in } \Omega_\ell, \\ u_\ell = 0 \text{ on } (-\ell, \ell) \times \partial\omega, \quad u_\ell = g \text{ on } \{-\ell, \ell\} \times \omega. \end{cases} \tag{2.73}$$

Assuming for instance g smooth we have

$$\begin{cases} -\Delta(u_\ell - g) = f + \Delta g = f + \partial_{x_2}^2 g & \text{in } \Omega_\ell, \\ u_\ell - g \in H_0^1(\Omega_\ell). \end{cases} \tag{2.74}$$

Moreover we have

$$\begin{cases} -\partial_{x_2}^2(u_\infty - g) = f + \partial_{x_2}^2 g & \text{in } \omega, \\ u_\infty - g \in H_0^1(\omega). \end{cases} \tag{2.75}$$

Thus applying Theorem 2.1 or 2.11 it holds that for any $\ell_0 > 1$

$$(u_\ell - g) - (u_\infty - g) = u_\ell - u_\infty \to 0 \quad \text{in } H^1(\Omega_{\ell_0}) \tag{2.76}$$

which shows that what matters really are the lateral homogeneous boundary conditions in order to obtain convergence toward u_∞.

The convergence in $H^1(\Omega_{\ell_0})$ is certainly well adapted to the structure of the problems. However, for instance when we come to the point of computations – and when we are planning to replace the computation of u_ℓ by the determination of u_∞ – it is important to have pointwise estimates. For instance it is important to show that u_ℓ converges toward u_∞ uniformly. In order to achieve that we can estimate $u_\ell - u_\infty$ in higher order Sobolev spaces. These estimates are inspired from similar estimates in regularity theory for instance. Let us explain that in the simple problem (2.1).

We have

THEOREM 2.12. $\forall \ell_0 > 0$, $\forall r > 0$ there exists a constant C independent of ℓ such that

$$|u_\ell - u_\infty|_{H^2(\Omega_{\ell_0})} \leq \frac{C}{\ell^r}. \tag{2.77}$$

PROOF. We have

$$\begin{cases} -\Delta(u_\ell - u_\infty) = 0 & \text{in } \Omega_\ell, \\ u_\ell - u_\infty = 0 & \text{on } (-\ell, \ell) \times \partial\omega. \end{cases} \tag{2.78}$$

From well-known regularity results – see for instance [19] – it follows that $u_\ell - u_\infty$ is smooth in Ω_ℓ. Differentiating the equation with respect to x_1,

$$-\Delta \partial_{x_1} u_\ell = 0 \quad \text{in } \Omega_\ell. \tag{2.79}$$

Let us denote by η a smooth function of x_1 such that

$$0 \le \eta \le 1, \quad \eta = 1 \text{ on } \Omega_{\ell_0}, \quad \eta = 0 \text{ in } \Omega_\ell \setminus \Omega_{\ell_0+1} \quad (\ell > \ell_0 + 1). \qquad (2.80)$$

Since $\partial_{x_1} u_\ell = 0$ on $(-\ell, \ell) \times \partial\omega$ the function

$$(\partial_{x_1} u_\ell)\eta^2 \in H_0^1(\Omega_\ell).$$

From (2.79) we derive, setting $w_\ell = \partial_{x_1} u_\ell$,

$$\int_{\Omega_\ell} \nabla w_\ell \cdot \nabla(w_\ell \eta^2) \, dx = 0$$

$$\Rightarrow \quad \int_{\Omega_\ell} \nabla w_\ell \{\nabla(w_\ell \eta)\eta + w_\ell \eta \nabla \eta\} \, dx = 0 \qquad (2.81)$$

$$\Rightarrow \quad \int_{\Omega_\ell} \nabla(w_\ell \eta)\nabla w_\ell \eta \, dx + \int_{\Omega_\ell} \eta \nabla w_\ell \nabla \eta w_\ell \, dx = 0.$$

Noticing that

$$\eta \nabla w_\ell = \nabla(w_\ell \eta) - w_\ell \nabla \eta$$

we obtain

$$\int_{\Omega_\ell} |\nabla(w_\ell \eta)|^2 \, dx = \int_{\Omega_\ell} w_\ell^2 |\nabla \eta|^2 \, dx. \qquad (2.82)$$

Thus we get

$$\int_{\Omega_{\ell_0}} |\nabla w_\ell|^2 \, dx \le C \int_{\Omega_{\ell_0+1}} w_\ell^2 \, dx.$$

Replacing w_ℓ by its value we derive

$$\int_{\Omega_{\ell_0}} (\partial_{x_1}^2 u_\ell)^2 + (\partial_{x_1}\partial_{x_2} u_\ell)^2 \, dx \le C \int_{\Omega_{\ell_0+1}} (\partial_{x_1} u_\ell)^2 \, dx \le \frac{C}{\ell^{2r}} \qquad (2.83)$$

for some constant C independent of ℓ – see (2.71).

On the other hand

$$-\Delta u_\ell = -\Delta u_\infty \quad \text{in } \Omega_\ell$$

which gives

$$-\partial_{x_2}^2 \{u_\ell - u_\infty\} = \partial_{x_1}^2 u_\ell. \qquad (2.84)$$

Combining (2.83), (2.84) it is clear that

$$\int_{\Omega_{\ell_0}} \partial_{x_1}^2(u_\ell - u_\infty)^2 + \partial_{x_1}\partial_{x_2}(u_\ell - u_\infty)^2 + \partial_{x_2}^2(u_\ell - u_\infty)^2 \, dx \le \frac{C}{\ell^{2r}}$$

and the result follows by (2.71). $\qquad \qquad \qquad \qquad \qquad \qquad \qquad \square$

It is clear that we can continue this process – we will come back later to this. Let us however explain briefly the idea. By (2.82), (2.83) taking now

$$w_\ell = \partial_{x_1}^2 u_\ell$$

we easily obtain as above

$$\int_{\Omega_{\ell_0}} (\partial_{x_1}^3 u_\ell)^2 + (\partial_{x_2}\partial_{x_1}^2 u_\ell)^2 \, dx \le \frac{C}{\ell^{2r}}$$

which is also

$$\int_{\Omega_{\ell_0}} \partial_{x_1}^3 (u_\ell - u_\infty)^2 + \partial_{x_2} \partial_{x_1}^2 (u_\ell - u_\infty)^2 \, dx \leq \frac{C}{\ell^{2r}}. \tag{2.85}$$

From the equation

$$-\Delta(u_\ell - u_\infty) = 0$$

we derive, taking the derivative in x_1,

$$\partial_{x_2}^2 \partial_{x_1} (u_\ell - u_\infty) = -\partial_{x_1}^3 (u_\ell - u_\infty) \tag{2.86}$$

and in x_2

$$\partial_{x_2}^3 (u_\ell - u_\infty) = -\partial_{x_2} \partial_{x_1}^2 (u_\ell - u_\infty). \tag{2.87}$$

It follows that

$$|u_\ell - u_\infty|_{H^3(\Omega_{\ell_0})} \leq \frac{C}{\ell^r}. \tag{2.88}$$

In particular since

$$H^3(\Omega_{\ell_0}) \hookrightarrow C(\overline{\Omega}_{\ell_0}),$$

where $C(\overline{\Omega}_{\ell_0})$ is the space of continuous functions on $\overline{\Omega}_{\ell_0}$ we have obtained

THEOREM 2.13. $\forall \ell_0 > 0$, $\forall r > 0$ there exists a constant C independent of ℓ such that

$$|u_\ell - u_\infty|_{\infty, \Omega_{\ell_0}} \leq \frac{C}{\ell^r}. \tag{2.89}$$

REMARK 2.4. This theorem justifies completely computing u_∞ instead of u_ℓ in a situation where u_ℓ is defined on a cylinder. Note that in the above arguments f played no role since it was turned around by the use of u_∞. Of course the Theorems 2.12, 2.13 hold in the case where u_ℓ is the solution corresponding to Ω_ℓ or Ω'_ℓ since the estimates in Ω_{ℓ_0} hold in both cases.

2.4. A different point of view

Following an idea of [31], we consider in this section a generalisation of the problems that we are addressing. However, in the following chapters, we will come back to our original point of view limiting our results in this new direction to this section.

Let $\Omega_\ell = (-\ell, \ell) \times (-1, 1)$ be the rectangle introduced in Figure 2.1 with $\omega = (-1, 1)$. We denote by Γ_0 the part of $\partial \Omega_\ell$ defined by

$$\Gamma_0 = (-\ell, \ell) \times \{-1, 1\} \tag{2.90}$$

and we consider V_ℓ^1, V_ℓ^2 two closed subspaces of $H_0^1(\Omega_\ell; \Gamma_0)$ such that

$$H_0^1(\Omega_\ell) \subset V_\ell^1, V_\ell^2 \subset H_0^1(\Omega_\ell; \Gamma_0). \tag{2.91}$$

Let $f_\ell = f_\ell(x_1, x_2)$ be an element of $H^1_0(\Omega_\ell; \Gamma_0)'$ the dual of $H^1_0(\Omega_\ell; \Gamma_0)$. Clearly the restriction of f_ℓ to V^i_ℓ defines a continuous linear form on V^i_ℓ, $i = 1, 2$. Thus for $i = 1, 2$ there exists a weak solution to

$$\begin{cases} u^i_\ell \in V^i_\ell, \\ \displaystyle\int_{\Omega_\ell} \nabla u^i_\ell \nabla v \, dx = \langle f_\ell, v \rangle \quad \forall v \in V^i_\ell. \end{cases} \tag{2.92}$$

In the above system $\langle \cdot \rangle$ denotes the duality bracket between $H^1_0(\Omega_\ell; \Gamma_0)'$ and $H^1_0(\Omega_\ell; \Gamma_0)$. Suppose now that we equipped $H^1_0(\Omega_\ell; \Gamma_0)$ with the norm

$$||\nabla v||_{2,\Omega_\ell} = \left\{ \int_{\Omega_\ell} |\nabla v(x)|^2 \, dx \right\}^{\frac{1}{2}}. \tag{2.93}$$

We denote then $|\cdot|_*$ the strong dual norm in $H^1_0(\Omega_\ell; \Gamma_0)'$. Then we have

THEOREM 2.14. *Suppose that there exists $\alpha > 0$, $C > 0$ constants independent of ℓ such that*

$$|f_\ell|_* \le C\ell^\alpha, \tag{2.94}$$

then $\forall \ell_0 > 0$, $\forall r > 0$ there exists a constant C independent of ℓ such that

$$||\nabla(u^1_\ell - u^2_\ell)||_{2,\Omega_{\ell_0}} \le \frac{C}{\ell^r} \tag{2.95}$$

i.e., $u^1_\ell - u^2_\ell \to 0$ in $H^1(\Omega_{\ell_0})$ with a speed larger than any power of $\frac{1}{\ell}$.

REMARK 2.5. In other words on $\{-\ell, \ell\} \times \omega$ we can consider for u_ℓ any boundary condition and it will not change the behaviour of u_ℓ on any subdomain Ω_{ℓ_0}. For instance we can choose

$$V^1_\ell = H^1_0(\Omega_\ell), \qquad V^2_\ell = H^1_0(\Omega_\ell; \Gamma_0)$$

i.e., u^1_ℓ is the solution of the Dirichlet problem in Ω_ℓ, u^2_ℓ the solution of a mixed problem with Neumann homogeneous boundary conditions on $\{-\ell, \ell\} \times \omega$. Then it holds that $u^1_\ell - u^2_\ell \to 0$ in $H^1(\Omega_{\ell_0})$ $\forall \ell_0 > 0$. In the particular case where $f_\ell = f(x_2)$, then clearly $u^2_\ell = u_\infty$ where u_∞ is the solution to (2.3). We have thus generalized Theorem 2.1.

PROOF OF THEOREM 2.14. Due to the embedding (2.91), we derive from (2.92) that

$$\int_{\Omega_\ell} \nabla(u^1_\ell - u^2_\ell)\nabla v \, dx = 0 \quad \forall v \in H^1_0(\Omega_\ell). \tag{2.96}$$

If ϱ is the function of Figure 2.2 for $\ell_1 \le \ell$ we have

$$v = (u^1_\ell - u^2_\ell)\varrho^2\left(\frac{x_1}{\ell_1}\right) \in H^1_0(\Omega_\ell)$$

and from (2.96) we derive

$$\int_{\Omega_\ell} |\nabla(u_\ell^1 - u_\ell^2)|^2 \varrho^2\left(\frac{x_1}{\ell_1}\right) dx$$
$$= -\frac{2}{\ell_1} \int_{\Omega_\ell} \partial_{x_1}(u_\ell^1 - u_\ell^2) \partial_{x_1}\varrho\left(\frac{x_1}{\ell_1}\right)(u_\ell^1 - u_\ell^2)\varrho\left(\frac{x_1}{\ell_1}\right) dx. \tag{2.97}$$

From the fact that $|\partial_{x_1}\varrho| \leq 2$ and Lemma 2.2 we get easily

$$\int_{\Omega_{\ell_1}} |\nabla(u_\ell^1 - u_\ell^2)|^2 \varrho^2\left(\frac{x_1}{\ell_1}\right) dx$$
$$\leq \frac{4}{\ell_1} \int_{\Omega_{\ell_1}} |\partial_{x_1}(u_\ell^1 - u_\ell^2)||(u_\ell^1 - u_\ell^2)|\varrho\left(\frac{x_1}{\ell_1}\right)$$
$$\leq \frac{4}{\ell_1} |\partial_{x_1}(u_\ell^1 - u_\ell^2)|_{2,\Omega_{\ell_1}} \left|(u_\ell^1 - u_\ell^2)\varrho\left(\frac{x_1}{\ell_1}\right)\right|_{2,\Omega_{\ell_1}} \tag{2.98}$$
$$\leq \frac{4\sqrt{2}}{\ell_1} |\partial_{x_1}(u_\ell^1 - u_\ell^2)|_{2,\Omega_{\ell_1}} \left|\partial_{x_2}(u_\ell^1 - u_\ell^2)\varrho\left(\frac{x_1}{\ell_1}\right)\right|_{2,\Omega_{\ell_1}}$$
$$\leq \frac{4\sqrt{2}}{\ell_1} \left|\left|\nabla(u_\ell^1 - u_\ell^2)\right|\right|_{2,\Omega_{\ell_1}} \left|\left|\nabla(u_\ell^1 - u_\ell^2)\right|\varrho\left(\frac{x_1}{\ell_1}\right)\right|_{2,\Omega_{\ell_1}}.$$

From this it follows that

$$\left|\left|\nabla(u_\ell^1 - u_\ell^2)\right|\varrho\left(\frac{x_1}{\ell}\right)\right|_{2,\Omega_{\ell_1}} \leq \frac{4\sqrt{2}}{\ell_1} \left|\left|\nabla(u_\ell^1 - u_\ell^2)\right|\right|_{2,\Omega_{\ell_1}}.$$

Since $\varrho(\frac{x_1}{\ell_1}) = 1$ on $\Omega_{\ell_1/2}$ this implies

$$\left|\left|\nabla(u_\ell^1 - u_\ell^2)\right|\right|_{2,\Omega_{\ell_1/2}} \leq \frac{4\sqrt{2}}{\ell_1} \left|\left|\nabla(u_\ell^1 - u_\ell^2)\right|\right|_{2,\Omega_{\ell_1}}.$$

Choosing $\ell_1 = \ell/2^{k-1}$ and iterating the above inequality we obtain

$$\left|\left|\nabla(u_\ell^1 - u_\ell^2)\right|\right|_{2,\Omega_{\ell/2^k}} \leq \frac{C_k}{\ell^k} \left|\left|\nabla(u_\ell^1 - u_\ell^2)\right|\right|_{2,\Omega_\ell} \tag{2.99}$$

where C_k is a constant independent of ℓ. Thus it remains – as before – to estimate the right-hand side of (2.99). For that, choosing $v = u_\ell^i$ in (2.92) we obtain for $i = 1, 2$

$$||\nabla u_\ell^i||_{2,\Omega_\ell}^2 = \langle f_\ell, u_\ell^i \rangle \leq |f_\ell|_* ||\nabla u_\ell^i||_{2,\Omega_\ell}.$$

Thus by (2.94) it follows that

$$||\nabla u_\ell^i||_{2,\Omega_\ell} \leq C\ell^\alpha \quad i = 1, 2 \tag{2.100}$$

where C is a constant independent of ℓ. Combining (2.99), (2.100) we get

$$\left|\left|\nabla(u_\ell^1 - u_\ell^2)\right|\right|_{2,\Omega_{\ell/2^k}} \leq \frac{C}{\ell^{k-\alpha}}$$

where C is a constant independent of ℓ. Choosing then k such that $k - \alpha > r$ and then ℓ such that $\ell_0 \leq \frac{\ell}{2^k}$, the result follows. This completes the proof of Theorem 2.14. □

Open problems

1. Is it possible to remove any assumption on ℓ' in Theorem 2.11? (See also Theorem 3.5).

Chapter 3

A General Asymptotic Theory for Linear Elliptic Problems

In this section we would like to develop a general theory of convergence for linear problems with non-constant coefficients when one or several directions go to $+\infty$. More precisely let us denote by Ω_ℓ the cylinder of $\mathbb{R}^n$ defined by

$$\Omega_\ell = (-\ell, \ell)^p \times \omega \tag{3.1}$$

where

$$\omega \text{ is a Lipschitz domain of } \mathbb{R}^{n-p}. \tag{3.2}$$

For $x = (x_1, \ldots, x_p, x_{p+1}, \ldots, x_n) \in \mathbb{R}^n$ we will set

$$X_1 = (x_1, \ldots, x_p), \qquad X_2 = (x_{p+1}, \ldots, x_n).$$

Then, for f a function or distribution depending on X_2 only, a_{ij}, a some functions, we can consider u_ℓ the weak solution to

$$\begin{cases} -\partial_{x_i}(a_{ij}(x)\partial_{x_j}u_\ell) + a(x)u_\ell = f & \text{in } \Omega_\ell, \\ u_\ell = 0 & \text{on } \partial\Omega_\ell. \end{cases}$$

(In the above system we adopt the summation convention in i, j.) If for $i, j = p + 1, \ldots, n$, $a_{ij} = a_{ij}(X_2)$ and also $a = a(X_2)$, we can introduce u_∞ the solution to

$$\begin{cases} -\displaystyle\sum_{i,j=p+1}^{n} \partial_{x_i}(a_{ij}(X_2)\partial_{x_j}u_\infty) + a(X_2)u_\infty = f & \text{in } \omega, \\ u_\infty = 0 & \text{on } \partial\omega, \end{cases}$$

where $\partial\omega$ denotes the boundary of ω. What we would like to analyze here is the convergence of u_ℓ toward u_∞ when $\ell \to +\infty$. For the sake of completeness we will consider more general boundary conditions. Let us make our assumptions more precise. For $i, j = 1, \ldots, n$ we denote by $a_{ij} = a_{ij}(x)$ functions such that

$$a_{ij} \in L^\infty(\mathbb{R}^p \times \omega), \qquad |a_{ij}|_\infty \le \Lambda \tag{3.3}$$

for some $\Lambda > 0$ ($|\cdot|_\infty$ denotes the usual $L^\infty(\mathbb{R}^p \times \omega)$-norm). Moreover, we assume that

$$a_{ij}(x) = a_{ij}(X_2) \quad \forall i = 1, \ldots, n, \quad \forall j = p + 1, \ldots, n, \tag{3.4}$$

i.e., the last $n - p$ columns of the matrix (a_{ij}) are independent of X_1. We suppose also that the matrix (a_{ij}) is uniformly elliptic in the sense that it holds for some $\lambda > 0$ that

$$a_{ij}(x)\xi_i\xi_j \geq \lambda|\xi|^2 \quad \forall\,\xi \in \mathbb{R}^n, \quad \text{a.e. } x \in \mathbb{R}^p \times \omega. \tag{3.5}$$

Let us also introduce a function a such that

$$a = a(X_2) \in L^\infty(\omega), \qquad 0 \leq a(X_2) \leq \Lambda \quad \text{a.e. } X_2 \in \omega. \tag{3.6}$$

Then, for $u, v \in H^1(\Omega_\ell)$, we will set, with the summation convention,

$$a(u, v) = \int_{\Omega_\ell} a_{ij}\partial_{x_j}u\partial_{x_i}v + auv\,dx. \tag{3.7}$$

Let $\partial_0\omega$ denote a measurable subset of $\partial\omega$ the boundary of ω. We will assume that

$$|\partial_0\omega| + a > 0 \tag{3.8}$$

when $|\cdot|$ denotes the measure area on $\partial\omega$. In other words when $|\partial_0\omega| = 0$ we suppose

$$a > 0 \text{ on a set of positive measure of } \omega. \tag{3.9}$$

For $u, v \in H^1(\omega)$ we define

$$a_\omega(u, v) = \int_\omega \sum_{i,j=p+1}^n a_{ij}(X_2)\partial_{x_j}u\partial_{x_i}v + a(X_2)uv\,dX_2. \tag{3.10}$$

In the above formula and subsequently dX_2 denotes the measure $dx_{p+1}\dots dx_n$. Then for $f \in H_0^1(\omega; \partial_0\omega)'$ – recall that $H_0^1(\omega; \partial_0\omega)$ is the set of functions of $H^1(\omega)$ vanishing on $\partial_0\omega$ in the sense explained in Chapter 1 – we introduce u_∞ the solution to

$$\begin{cases} a_\omega(u_\infty, v) = \langle f, v \rangle \quad \forall\,v \in H_0^1(\omega; \partial_0\omega), \\ u_\infty \in H_0^1(\omega; \partial_0\omega). \end{cases} \tag{3.11}$$

We define then

$$\Gamma_0 = \partial(-\ell, \ell)^p \times \omega \cup (-\ell, \ell)^p \times \partial_0\omega, \tag{3.12}$$

where $\partial(-\ell, \ell)^p$ denotes the boundary of $(-\ell, \ell)^p$.

Let us first prove the auxilliary proposition:

PROPOSITION 3.1. *Let* $v \in H_0^1(\Omega_\ell; \Gamma_0)$ – *see* (1.28), (1.36). *Then, it holds that*

$$v(X_1, \cdot) \in H_0^1(\omega; \partial_0\omega) \quad \text{a.e. } X_1 \in (-\ell, \ell)^p. \tag{3.13}$$

PROOF. Let $v \in H_0^1(\Omega_\ell; \Gamma_0)$. By (1.28) there exists a sequence of functions (v_n) such that $v_n \in C_0^1(\overline{\Omega}_\ell; \Gamma_0)$ and

$$\int_{\Omega_\ell} |\nabla(v - v_n)|^2 + |v_n - v|^2\,dx \to 0. \tag{3.14}$$

It follows that – up to a subsequence – we have

$$\int_\omega |\nabla_{X_2}(v - v_n)(X_1, X_2)|^2 + |(v_n - v)(X_1, X_2)|^2 \, dX_2 \to 0 \qquad (3.15)$$

for a.e. $X_1 \in (-\ell, \ell)^p$ – (∇_{X_2} denotes the vector of components $(\partial_{x_{p+1}}, \ldots, \partial_{x_n})$.)
Thus for a.e. $X_1 \in (-\ell, \ell)^p$, (3.13) holds – recall that $\nabla_{X_2} v(X_1, \cdot)$ is also the vector
of the derivatives of $v(X_1, \cdot)$ in the distributional sense – see for instance below
(2.50) for a similar argument. This completes the proof of the proposition. $\square$

Then, for $f \in (H_0^1(\omega; \partial_0\omega))'$, $v \in H_0^1(\Omega_\ell; \Gamma_0)$ we can define

$$\langle f, v(X_1, \cdot) \rangle. \qquad (3.16)$$

On $V = H_0^1(\omega; \partial_0\omega)$ and $V_{\Omega_\ell} = H_0^1(\Omega_\ell; \Gamma_0)$ we will consider the norms

$$|v|_V = |v|_{a,\omega} = \left\{ \int_\omega |\nabla_{X_2} v|^2 + av^2 \, dx \right\}^{1/2}, \qquad (3.17)$$

$$|v|_{V_{\Omega_\ell}} = |v|_{a,\Omega_\ell} = \left\{ \int_{\Omega_\ell} |\nabla v|^2 + av^2 \, dx \right\}^{1/2}, \qquad (3.18)$$

that are adapted to the problem. If $| \cdot |_*$ denotes the strong dual norm in V' the
dual of V, we have then

$$. \; |\langle f_1, v(X_1, \cdot) \rangle| \leq |f|_* |v(X_1, \cdot)|_{a,\omega}. \qquad (3.19)$$

It follows that the function of X_1 defined by (3.16) is measurable in X_1. Indeed,
if v_n is a sequence of smooth functions approximating v, then for X_1, X_1' it holds
that

$$|\langle f, v_n(X_1, \cdot) \rangle - \langle f, v_n(X_1', \cdot) \rangle| \leq |f|_* |v_n(X_1, \cdot) - v_n(X_1', \cdot)|_{a,\omega}$$
$$\leq \varepsilon \quad \text{for } X_1, X_1' \text{ close.}$$

This shows that

$$X_1 \mapsto \langle f, v_n(X_1, \cdot) \rangle$$

is continuous. From (3.19) and Proposition 3.1 we have also pointwise

$$\langle f, v_n(X_1, \cdot) \rangle \to \langle f, v(X_1, \cdot) \rangle \quad \text{a.e. } X_1$$

and thus the function defined by (3.16) is measurable. It holds that

$$\int_{(-\ell,\ell)^p} |\langle f, v(X_1, \cdot) \rangle| \, dx \leq \int_{(-\ell,\ell)^p} |f|_* |v(X_1, \cdot)|_{a,\omega} \, dX_1$$

$$\leq |f|_* \int_{(-\ell,\ell)^p} |v(X_1, \cdot)|_{a,\omega} \, dX_1$$

$$\leq |f|_* (2\ell)^{\frac{p}{2}} \left\{ \int_{(-\ell,\ell)^p} |v(X_1, \cdot)|_{a,\omega}^2 \, dX_1 \right\}^{\frac{1}{2}} \qquad (3.20)$$

$$\leq (2\ell)^{\frac{p}{2}} |f|_* \left\{ \int_{\Omega_\ell} |\nabla_{X_2} v|^2 + av^2 \, dx \right\}^{\frac{1}{2}} < +\infty.$$

It follows that the function defined in (3.16) is integrable. Thus for $v \in H^1_0(\Omega_\ell; \Gamma_0)$ we will set

$$\langle \tilde{f}, v \rangle = \int_{(-\ell,\ell)^p} \langle f, v(\mathrm{x}_1, \cdot) \rangle \, d\mathrm{x}_1. \tag{3.21}$$

Clearly $\tilde{f}$ is a continuous linear form on $V_{\Omega_\ell} = H^1_0(\Omega_\ell, \Gamma_0)$ – see (3.20). Since there is no ambiguity we will denote it also by f.

REMARK 3.1. For instance when f is a function $f(\mathrm{x}_2)$ in $L^2(\omega)$, then (3.21) can be written as

$$\langle \tilde{f}, v \rangle = \int_{(-\ell,\ell)} \langle f, v(\mathrm{x}_1, \cdot) \rangle \, d\mathrm{x}_1 = \int_{(-\ell,\ell)} \int_\omega f(\mathrm{x}_2) v(\mathrm{x}_1, \mathrm{x}_2) \, d\mathrm{x}_2 \, d\mathrm{x}_1$$
$$= \int_{\Omega_\ell} f(\mathrm{x}_2) v(x) \, dx \tag{3.22}$$

and $\tilde{f}$ corresponds really to f.

Having defined (3.21) we can introduce u_ℓ the solution to

$$\begin{cases} \int_{\Omega_\ell} a_{ij}(x) \partial_{x_j} u_\ell \partial_{x_i} v + a u_\ell v \, dx = \langle f, v \rangle \quad \forall v \in V_{\Omega_\ell}, \\ u_\ell \in V_{\Omega_\ell}. \end{cases} \tag{3.23}$$

To fix the ideas recall that

$$V_{\Omega_\ell} = H^1_0(\Omega_\ell; \Gamma_0)$$

and thus the boundary conditions are

$$u_\ell = 0 \quad \text{on} \quad \partial(-\ell, \ell)^p \times \omega \quad \text{and} \quad (-\ell, \ell)^p \times \partial_0 \omega.$$

3.1. A general convergence result in $H^1(\Omega_{\ell_0})$

In this section we would like to extend the model results of Chapter 2 to the general situation above. It will be convenient to adopt the following notation. If A denotes the matrix $(a_{ij})_{i,j=1,\dots,n}$ we will set

$$A = \begin{pmatrix} A_{11} & A_{12} \\ A_{21} & A_{22} \end{pmatrix} \tag{3.24}$$

where A_{ij} $i = 1, 2$ are the submatrices defined by

$$A_{11} = (a_{ij})_{i,j=1,\dots,p}, \qquad A_{12} = (a_{ij})_{\substack{i=1,\dots,p \\ j=p+1,\dots,n}}, \tag{3.25}$$

$$A_{21} = (a_{ij})_{\substack{i=p+1,\dots,n \\ j=1,\dots,p}}, \qquad A_{22} = (a_{ij})_{i,j=p+1,\dots,n}. \tag{3.26}$$

Then let us prove

THEOREM 3.2. *Set* $V = H_0^1(\omega; \partial_0\omega)$, $V_{\Omega_\ell} = H_0^1(\Omega_\ell; \Gamma_0)$. *Under the above assumptions, i.e.,* (3.1)–(3.12), *for* $f \in V'$, *let* u_ℓ *be the solution to* (3.23) *and* u_∞ *the solution to* (3.11). *Then, for any* $\ell_0 > 0$ *and any* $r > 0$ *there exists a constant* C *independent of* ℓ *such that*

$$|u_\ell - u_\infty|_{V_{\Omega_{\ell_0}}} \leq \frac{C}{\ell^r}. \tag{3.27}$$

(| $|_{V_{\Omega_{\ell_0}}} = | \ |_{a,\Omega_{\ell_0}}$ is the norm defined by (3.18).)

The proof goes through the three steps described in Chapter 2. First we have

LEMMA 3.3. *For any* $v \in V_{\Omega_\ell}$ *it holds that*

$$|v|_{2,\Omega_\ell} \leq C \left\{ \int_{\Omega_\ell} |\nabla_{X_2} v|^2 + av^2 \right\}^{\frac{1}{2}} \tag{3.28}$$

where C *is a constant independent of* ℓ *and* $\nabla_{X_2} v$ *the vector* $(\partial_{x_{p+1}} v, \ldots, \partial_{x_n} v)$, $|\cdot|$ *the euclidean norm in* $\mathbb{R}^{n-p}$.

PROOF. Applying Theorems 1.3, 1.5 and Proposition 3.1 it holds that for some constant C

$$\int_\omega v(X_1, X_2)^2 \, dX_2 \leq C \int_\omega |\nabla_{X_2} v(X_1, X_2)|^2 + av^2(X_1, X_2) \, dX_2$$

where C is a constant independent of v. Integrating on $(-\ell, \ell)^p$ we obtain

$$\int_{\Omega_\ell} v^2 \, dx \leq C \int_{\Omega_\ell} |\nabla_{X_2} v|^2 + av^2 \, dx.$$

This completes the proof of the Lemma.   □

REMARK 3.2. In the case $|\partial_0\omega| > 0$ the function a could be taken equal to 0 in (3.28). In the case where $|\partial_0\omega| = 0$ we suppose $a > 0$ on a set of positive measure and $H_0^1(\omega; \partial_0\omega)$ is simply the space $H^1(\omega)$. The notation is of course not very satisfactory but we can treat in one single theory the case of Dirichlet, mixed and Neumann boundary conditions for u_∞.

We can now estimate u_ℓ. We have

LEMMA 3.4. *Under the assumptions of Theorem 3.2 it holds that*

$$|u_\ell|_{a,\Omega_\ell} \leq C\ell^{p/2} |u_\infty|_{a,\omega} \tag{3.29}$$

where $C = c(\lambda, \Lambda, n)$ *is independent of* ℓ.

PROOF. From (3.11), (3.21), (3.23) it holds that

$$\int_{\Omega_\ell} A\nabla u_\ell \nabla v + au_\ell v \, dx = \int_{\Omega_\ell} A_{22} \nabla_{X_2} u_\infty \nabla_{X_2} v + au_\infty v \, dx \quad \forall \, v \in V_{\Omega_\ell}. \tag{3.30}$$

($A\nabla u_\ell$ denotes the multiplication of the matrix A by the vector ∇u_ℓ; $A_{22}\nabla_{X_2}u_\infty$ the multiplication of A_{22} by $\nabla_{X_2}u_\infty$.) Taking $v = u_\ell$ we obtain

$$\int_{\Omega_\ell} A\nabla u_\ell \nabla u_\ell + a u_\ell^2 \, dx = \int_{\Omega_\ell} A_{22}\nabla_{X_2}u_\infty \nabla_{X_2}u_\ell + a u_\infty u_\ell \, dx.$$

By (3.3), (3.5) it follows that

$$\int_{\Omega_\ell} \lambda |\nabla u_\ell|^2 + a u_\ell^2 \, dx$$

$$\leq C \int_{\Omega_\ell} |\nabla_{X_2}u_\infty||\nabla_{X_2}u_\ell| + a|u_\infty u_\ell| \, dx \qquad (3.31)$$

$$\leq C \left\{ \int_{\Omega_\ell} |\nabla_{X_2}u_\infty|^2 + a u_\infty^2 \, dx \right\}^{\frac{1}{2}} \left\{ \int_{\Omega_\ell} |\nabla_{X_2}u_\ell|^2 + a u_\ell^2 \, dx \right\}^{\frac{1}{2}}.$$

(We used the Cauchy–Schwarz inequality for the form appearing in the right-hand side of (3.31).) It follows that

$$\min(\lambda, 1)|u_\ell|_{a,\Omega_\ell}^2 \leq C(2\ell)^{p/2}|u_\infty|_{a,\omega}|u_\ell|_{a,\Omega_\ell}$$

which clearly implies (3.29) after having divided the above inequality through $|u_\ell|_{a,\Omega_\ell}$. $\qquad\square$

We can now complete the proof of Theorem 3.2.

PROOF OF THEOREM 3.2. Let us denote by ϱ a smooth function of $\mathbb{R}^p$ such that

$$0 \leq \varrho \leq 1, \quad \varrho = 1 \quad \text{on} \quad \left(-\frac{1}{2}, \frac{1}{2}\right)^p, \quad \varrho = 0 \quad \text{on} \quad \mathbb{R}^p \setminus (-1,1)^p, \quad (3.32)$$

$$|\nabla_{X_1}\varrho| \leq \theta \qquad (3.33)$$

where θ is some constant – recall that $\nabla_{X_1}\varrho$ denotes the gradient of ϱ i.e., $\nabla_{X_1}\varrho = (\partial x_1\varrho, \ldots, \partial x_p\varrho)$. Clearly, for every $\ell_1 \leq \ell$

$$(u_\ell - u_\infty)\varrho^2\left(\frac{X_1}{\ell_1}\right) \in V_{\Omega_\ell}.$$

Thus from (3.30) we derive

$$\int_{\Omega_\ell} A\nabla u_\ell \nabla\left\{(u_\ell - u_\infty)\varrho^2\left(\frac{X_1}{\ell_1}\right)\right\} + a u_\ell(u_\ell - u_\infty)\varrho^2\left(\frac{X_1}{\ell}\right) dx$$

$$= \int_{\Omega_\ell} A_{22}\nabla_{X_2}u_\infty \nabla_{X_2}(u_\ell - u_\infty)\varrho^2\left(\frac{X_1}{\ell_1}\right) \qquad (3.34)$$

$$+ a u_\infty(u_\ell - u_\infty)\varrho^2\left(\frac{X_1}{\ell_1}\right) dx.$$

(On the right-hand side of (3.34) we used the fact that $\varrho(\frac{X_1}{\ell_1})$ is independent of X_2.) Using the decomposition of A in submatrices we have

$$A\nabla u \cdot \nabla v = \begin{pmatrix} A_{11} & A_{12} \\ A_{21} & A_{22} \end{pmatrix} \begin{pmatrix} \nabla_{X_1} u_\ell \\ \nabla_{X_2} u_\ell \end{pmatrix} \cdot \begin{pmatrix} \nabla_{X_1} v \\ \nabla_{X_2} v \end{pmatrix}$$

$$= \begin{pmatrix} A_{11}\nabla_{X_1} u_\ell + A_{12}\nabla_{X_2} u_\ell \\ A_{21}\nabla_{X_1} u_\ell + A_{22}\nabla_{X_2} u_\ell \end{pmatrix} \cdot \begin{pmatrix} \nabla_{X_1} v \\ \nabla_{X_2} v \end{pmatrix} \tag{3.35}$$

and thus (3.34) becomes

$$\int_{\Omega_\ell} A_{11}\nabla_{X_1} u_\ell \nabla_{X_1} \left\{ (u_\ell - u_\infty)\varrho^2\left(\frac{X_1}{\ell_1}\right) \right\} dx$$

$$+ \int_{\Omega_\ell} A_{12}\nabla_{X_2} u_\ell \nabla_{X_1} \left\{ (u_\ell - u_\infty)\varrho^2\left(\frac{X_1}{\ell_1}\right) \right\} dx$$

$$+ \int_{\Omega_\ell} A_{21}\nabla_{X_1} u_\ell \nabla_{X_2} (u_\ell - u_\infty)\varrho^2\left(\frac{X_1}{\ell_1}\right) dx$$

$$+ \int_{\Omega_\ell} A_{22}\nabla_{X_2} (u_\ell - u_\infty)\nabla_{X_2} (u_\ell - u_\infty)\varrho^2\left(\frac{X_1}{\ell_1}\right) dx$$

$$+ \int_{\Omega_\ell} a(u_\ell - u_\infty)^2\varrho^2\left(\frac{X_1}{\ell_1}\right) dx = 0.$$

Performing the derivation in X_1 and using the independence of u_∞ with respect to X_1 we obtain

$$\sum_{i,j=1}^{2} \int_{\Omega_\ell} A_{ij} \cdot \nabla_{X_j}(u_\ell - u_\infty)\nabla_{X_i}(u_\ell - u_\infty)\varrho^2\left(\frac{X_1}{\ell_1}\right) dx$$

$$+ \int_{\Omega_\ell} a(u_\ell - u_\infty)^2\varrho^2\left(\frac{X_1}{\ell_1}\right) dx$$

$$= -\frac{2}{\ell_1} \int_{\Omega_\ell} A_{11}\nabla_{X_1}(u_\ell - u_\infty)\nabla_{X_1}\varrho\left(\frac{X_1}{\ell_1}\right)(u_\ell - u_\infty)\varrho\left(\frac{X_1}{\ell_1}\right) dx \tag{3.36}$$

$$- \frac{2}{\ell_1} \int_{\Omega_\ell} A_{12}\nabla_{X_2}(u_\ell - u_\infty)\nabla_{X_1}\varrho\left(\frac{X_1}{\ell_1}\right)(u_\ell - u_\infty)\varrho\left(\frac{X_1}{\ell_1}\right) dx$$

$$- \int_{\Omega_\ell} A_{12}\nabla_{X_2} u_\infty \nabla_{X_1} \left\{ (u_\ell - u_\infty)\varrho^2\left(\frac{X_1}{\ell_1}\right) \right\} dx.$$

We claim first that the last integral above is equal to 0. Indeed since A_{12} is independent of X_1 and $\nabla_{X_2} u_\infty$ also, it holds that

$$\int_{\Omega_\ell} A_{12}\nabla_{X_2} u_\infty \nabla_{X_1} \left\{ (u_\ell - u_\infty)\varrho^2\left(\frac{X_1}{\ell_1}\right) \right\} dx$$

$$= \int_{\Omega_\ell} \nabla_{X_1} \cdot \left\{ (u_\ell - u_\infty)\varrho^2\left(\frac{X_1}{\ell_1}\right) A_{12}\nabla_{X_2} u_\infty \right\} dx \tag{3.37}$$

$$= 0.$$

(We integrate in the X_1 direction first and use the fact that $\varrho(\frac{X_1}{\ell_1}) = 0$ outside $(-\ell_1, \ell_1)^p$.) Thus from (3.36) using the ellipticity condition (3.5), (3.3), (3.33) we obtain

$$\int_{\Omega_{\ell_1}} \{\lambda|\nabla(u_\ell - u_\infty)|^2 + a(u_\ell - u_\infty)^2\}\varrho^2\left(\frac{X_1}{\ell_1}\right)dx$$

$$\leq \frac{C}{\ell_1}\int_{\Omega_{\ell_1}} |\nabla(u_\ell - u_\infty)||u_\ell - u_\infty|\varrho\left(\frac{X_1}{\ell_1}\right)dx \tag{3.38}$$

where C is a constant depending on Λ and n. It follows that

$$\min(1, \lambda)\int_{\Omega_{\ell_1}} \{|\nabla(u_\ell - u_\infty)|^2 + a(u_\ell - u_\infty)^2\}\varrho^2\left(\frac{X_1}{\ell_1}\right)dx$$

$$\leq \frac{C}{\ell_1}\left\|\nabla(u_\ell - u_\infty)\right\|_{2,\Omega_{\ell_1}}\left|(u_\ell - u_\infty)\varrho\left(\frac{X_1}{\ell_1}\right)\right|_{2,\Omega_{\ell_1}}. \tag{3.39}$$

We apply then Lemma 3.3 (with $\ell = \ell_1$) to get for some constant $C = C(\lambda, \Lambda, n)$

$$\int_{\Omega_{\ell_1}} \{|\nabla(u_\ell - u_\infty)|^2 + a(u_\ell - u_\infty)^2\}\varrho^2\left(\frac{X_1}{\ell_1}\right)dx$$

$$\leq \frac{C}{\ell_1}\left\|\nabla(u_\ell - u_\infty)\right\|_{2,\Omega_{\ell_1}} \tag{3.40}$$

$$\times \left(\int_{\Omega_{\ell_1}} \{|\nabla_{X_2}(u_\ell - u_\infty)|^2 + a(u_\ell - u_\infty)^2\}\varrho^2\left(\frac{X_1}{\ell_1}\right)dx\right)^{\frac{1}{2}}.$$

It follows that

$$\left(\int_{\Omega_{\ell_1}} \{|\nabla(u_\ell - u_\infty)|^2 + a(u_\ell - u_\infty)^2\}\varrho^2\left(\frac{X_1}{\ell_1}\right)dx\right)^{\frac{1}{2}}$$

$$\leq \frac{C}{\ell_1}\left\|\nabla(u_\ell - u_\infty)\right\|_{2,\Omega_{\ell_1}}. \tag{3.41}$$

Since $\varrho = 1$ on $(-\frac{1}{2}, \frac{1}{2})^p$ we get

$$\int_{\Omega_{\ell_1/2}} |\nabla(u_\ell - u_\infty)|^2 + a(u_\ell - u_\infty)^2\, dx \leq \frac{C^2}{\ell_1^2}\int_{\Omega_{\ell_1}} |\nabla(u_\ell - u_\infty)|^2\, dx. \tag{3.42}$$

In particular we have obtained for every $\ell_1 \leq \ell$

$$\int_{\Omega_{\ell_1/2}} |\nabla(u_\ell - u_\infty)|^2\, dx \leq \frac{C^2}{\ell_1^2}\int_{\Omega_{\ell_1}} |\nabla(u_\ell - u_\infty)|^2\, dx. \tag{3.43}$$

Taking $\ell_1 = \ell/2^{k-2}$ and iterating the above formula we obtain

$$\int_{\Omega_{\ell/2^{k-1}}} |\nabla(u_\ell - u_\infty)|^2\, dx \leq \frac{C^2}{\ell^2}2^{2(k-2)}\int_{\Omega_{\ell/2^{k-2}}} |\nabla(u_\ell - u_\infty)|^2\, dx$$

$$\leq \frac{C^2}{\ell^{2(k-1)}}\int_{\Omega_\ell} |\nabla(u_\ell - u_\infty)|^2\, dx, \tag{3.44}$$

for some constant C independent of ℓ. Going back to (3.42) written for $\ell_1 = \frac{\ell}{2^{k-1}}$ we obtain

$$\int_{\Omega_{\ell/2k}} |\nabla(u_\ell - u_\infty)|^2 + a(u_\ell - u_\infty)^2 \, dx \leq \frac{C^2}{\ell^{2k}} \int_{\Omega_\ell} |\nabla(u_\ell - u_\infty)|^2 \, dx \qquad (3.45)$$

for some constant $C = C(\lambda, \Lambda, k, n)$. Using now Lemma 3.4 we derive

$$\int_{\Omega_\ell} |\nabla(u_\ell - u_\infty)|^2 \, dx \leq C\ell^p |u_\infty|_{a,\omega}^2$$

and thus we obtain

$$|u_\ell - u_\infty|_{a,\Omega_{\ell/2k}} \leq \frac{C}{\ell^{k-\frac{p}{2}}} |u_\infty|_{a,\omega}. \qquad (3.46)$$

We choose then $k - \frac{p}{2} > r$ and then $\frac{\ell}{2^k} > \ell_0$ to obtain

$$|u_\ell - u_\infty|_{a,\Omega_{\ell_0}} \leq \frac{C}{\ell^r} |u_\infty|_{a,\omega}$$

which completes the proof. $\qquad\qquad\qquad\qquad\qquad\qquad\qquad\qquad\qquad\qquad\square$

REMARK 3.3. In the above proof we can even allow some coefficients to depend on ℓ. Indeed if

$$A_{11}(x) = A_{11}^\ell(x), \qquad A_{21}(x) = A_{21}^\ell(x), \qquad A_{12}(\mathbf{x}_2) = A_{12}^\ell(\mathbf{x}_2) \qquad (3.47)$$

all the arguments developed above remain true provided we assume

$$A^\ell \xi\xi \geq \lambda |\xi|^2 \qquad (3.48)$$

$$|a_{ij}^\ell(x)| \leq \Lambda \quad \forall i,j = 1,\ldots,n, \qquad (3.49)$$

with λ, Λ independent of ℓ.

In Theorem 3.2 we could replace the space $H_0^1(\Omega_\ell; \Gamma_0)$ by any subspace V_ℓ such that

$$H_0^1(\Omega_\ell; \Gamma_0) \subset V_\ell \subset H_0^1(\Omega_\ell; \Gamma_0')$$

where $\Gamma_0' = (-\ell, \ell)^p \times \partial_0 \omega$.

We have shown that the rate of convergence of u_ℓ toward u_∞ can be a power of $\frac{1}{\ell}$ arbitrarily large. In the next section we are going to see that this rate can even be improved in some cases.

3.2. A sharper rate of convergence

In this section we assume the same conditions as above with in addition

$$A_{11} = A_{11}(\mathbf{x}_1), \qquad A_{22} = A_{22}(\mathbf{x}_2), \qquad A_{12} = A_{21} = 0, \qquad (3.50)$$

$$A_{22} \text{ is symmetric}, \qquad (3.51)$$

and for some constant D

$$|\partial_{x_k} a_{ij}(\mathbf{x}_1)| \leq D \quad \forall i,j,k = 1,\ldots,p. \qquad (3.52)$$

We consider then u_ℓ solution to

$$\begin{cases} \int_{\Omega_l} A_{11}\nabla_{X_1} u_\ell \nabla_{X_1} v + A_{22}\nabla_{X_2} u_\ell \nabla_{X_2} v + au_\ell v \, dx = \langle f, v \rangle \quad \forall v \in V_{\Omega_\ell}, \\ u_\ell \in V_{\Omega_\ell}, \end{cases} \tag{3.53}$$

and u_∞ solution to

$$\begin{cases} \int_\omega A_{22}\nabla_{X_2} u_\infty \nabla_{X_2} v + au_\infty v \, dX_2 = \langle f, v \rangle \quad \forall v \in V, \\ u_\infty \in V. \end{cases} \tag{3.54}$$

Recall that $V = H_0^1(\omega; \partial_0\omega)$, $V_{\Omega_\ell} = H_0^1(\Omega_\ell; \Gamma_0)$ with Γ_0 defined by (3.12), $f \in V'$ is extended to V_{Ω_ℓ} by (3.21). Then we would like to show

THEOREM 3.5. *Let us assume the conditions of Theorem 3.2 with in addition* (3.50)–(3.52). *Let u_ℓ, u_∞ be the solution to* (3.53)–(3.54) *respectively; then for every $\ell_0 > 0$ there exists two positive constants c_1, c_2 independent of ℓ such that*

$$|u_\ell - u_\infty|_{a,\Omega_{\ell_0}} \leq c_1 e^{-c_2\ell}. \tag{3.55}$$

(Note that as in Theorem 3.2, $u_l - u_\infty \notin V_{\Omega_{\ell_0}}$ since $u_l - u_\infty$ does not match the boundary conditions – however the quantity $|u_\ell - u_\infty|_{a,\Omega_{\ell_0}}$ is well defined.)

PROOF. Let $\lambda_1, \lambda_2, \ldots, \lambda_m \ldots$ be the eigenvalues of the problem whose operator is associated to (3.54). Let $\varphi_1, \ldots, \varphi_m \ldots$ be the corresponding normalized eigenfunctions. In other words the functions φ_m are the solutions to

$$\begin{cases} \int_\omega A_{22}\nabla_{X_2}\varphi_m\nabla_{X_2}v + a\varphi_m v \, dX_2 = \int_\omega \lambda_m\varphi_m v \, dX_2 \quad \forall v \in V, \\ \varphi_m \in V, \quad \int_\omega \varphi_m^2 \, dX_2 = 1. \end{cases} \tag{3.56}$$

It is well known that such a basis of orthonormalized eigenfunctions exists. Then for every λ_m (we know that $\lambda_m > 0$) we introduce $v_m = v_m(X_1)$ the solution to

$$\begin{cases} \int_{(-\ell,\ell)^p} A_{11}\nabla_{X_1}v_m\nabla_{X_1}v + \lambda_m v_m v \, dX_1 = 0 \quad \forall v \in H_0^1((-\ell,\ell)^p), \\ v_m = -1 \text{ on } \partial(-\ell,\ell)^p, \end{cases} \tag{3.57}$$

where $\partial(-\ell, \ell)^p$ denotes the boundary of $(-\ell, \ell)^p$. Clearly such v_m does exist. Then, it holds that

$$\int_{\Omega_\ell} A_{11}\nabla_{X_1}v_m\varphi_m\nabla_{X_1}v + A_{22}\nabla_{X_2}v_m\varphi_m\nabla_{X_2}v + av_m\varphi_m v \, dx$$

$$= \int_{(-\ell,\ell)^p}\int_\omega \varphi_m\{A_{11}\nabla_{X_1}v_m\nabla_{X_1}v\} + v_m\{A_{22}\nabla_{X_2}\varphi_m\nabla_{X_2}v + a\varphi_m v\} \, dx$$

$$= \int_\omega \varphi_m \int_{(-\ell,\ell)^p} A_{11} \nabla_{X_1} v_m \nabla_{X_1} v \, dx$$

$$+ \int_{(-\ell,\ell)^p} v_m \int_\omega A_{22} \nabla_{X_2} \varphi_m \nabla_{X_2} v + a\varphi_m v \, dx \tag{3.58}$$

$$= \int_{\Omega_\ell} -\lambda_m \varphi_m v_m v + \lambda_m \varphi_m v_m v \, dx = 0 \quad \forall v \in V_{\Omega_\ell}.$$

Since (φ_m) is an orthonormal basis of eigenfunctions it holds that

$$u_\infty = \sum_{m=1}^{+\infty} a_m \varphi_m \tag{3.59}$$

with

$$a_m = (u_\infty, \varphi_m)_2 \tag{3.60}$$

where $(\cdot, \cdot)_2$ is the usual scalar product in $L^2(\omega)$. The convergence in (3.59) is taking place in $H^1(\omega)$. From (3.53), (3.54) we derive that

$$\int_{\Omega_\ell} A_{11} \nabla_{X_1} (u_\ell - u_\infty) \nabla_{X_1} v + A_{22} \nabla_{X_2} (u_\ell - u_\infty) \nabla_{X_2} v + a(u_\ell - u_\infty)v \, dx = 0 \tag{3.61}$$

for any $v \in V_{\Omega_\ell}$. From (3.58), (3.59) we then deduce (we will justify the convergence later)

$$u_\ell - u_\infty = \sum_{m=1}^{+\infty} a_m v_m \varphi_m. \tag{3.62}$$

We know that

$$\lambda_m > 0, \qquad \lambda_m \uparrow \text{ with } m, \qquad \lim_{m \to +\infty} \lambda_m = +\infty. \tag{3.63}$$

Then we can show:

LEMMA 3.6. *It holds that*

$$-1 \leq v_m \leq 0, \qquad v_m \nearrow \text{ with } m. \tag{3.64}$$

PROOF. Taking as test function in (3.57), $v = v_m^+$ where $(\cdot)^+$ denotes the positive part of a function we obtain

$$\int_{(-\ell,\ell)^p} A_{11} \nabla_{X_1} v_m^+ \nabla_{X_1} v_m^+ + \lambda_m v_m^+ v_m^+ \, dX_1 = 0 \quad \Rightarrow \quad v_m \leq 0. \tag{3.65}$$

Next taking $v = (-1 - v_m)^+$ we obtain

$$\int_{(-\ell,\ell)} A_{11} \nabla_{X_1} v_m \nabla_{X_1} (-1 - v_m)^+ \, dX_1 = -\int_{(-\ell,\ell)^p} \lambda_m v_m (-1 - v_m)^+ \, dX_1 \geq 0.$$

Hence

$$\int_{(-l,l)^p} A_{11} \nabla_{X_1} (-1 - v_m)^+ \nabla_{X_1} (-1 - v_m)^+ \, dX_1 \leq 0 \tag{3.66}$$

and the inequality (3.64) holds. Next from (3.57) we have

$$\int_{(-l,l)^p} A_{11} \nabla_{X_1} (v_m - v_{m+1}) \nabla_{X_1} v + \lambda_m (v_m - v_{m+1}) v \, dX_1$$

$$= \int_{(-\ell,\ell)^p} (\lambda_{m+1} - \lambda_m) v_{m+1} v \, dX_1 \tag{3.67}$$

for every $v \in H_0^1((-\ell,\ell)^p)$. Taking $v = (v_m - v_{m+1})^+$ we derive

$$(v_m - v_{m+1})^+ = 0 \quad \Rightarrow \quad v_m \le v_{m+1} \tag{3.68}$$

which completes the proof of (3.64). $\qquad\square$

Due to the orthonormality of the φ_m's we have – see (3.56)

$$\int_\omega A_{22} \nabla_{X_2} \varphi_m \nabla_{X_2} \varphi_p + a \varphi_m \varphi_p \, dX_2 = \delta_{mp} \lambda_m \tag{3.69}$$

where δ_{mp} denotes the Konecker symbol.

Let us come back to (3.62) to show that the series introduced there converges in $H^1(\Omega_\ell)$. Set

$$S_n = \sum_{m=1}^n a_m \varphi_m v_m.$$

We have for $n > q$

$$\nabla_{X_1}(S_n - S_q) = \sum_{m=q+1}^n a_m \varphi_m \nabla_{X_1} v_m, \qquad \nabla_{X_2}(S_n - S_q) = \sum_{m=q+1}^n a_m v_m \nabla_{X_2} \varphi_m.$$

It follows that

$$\int_{\Omega_\ell} A_{11} \nabla_{X_1}(S_n - S_q) \nabla_{X_1}(S_n - S_q)$$

$$+ A_{22} \nabla_{X_2}(S_n - S_q) \nabla_{X_2}(S_n - S_q) + a(S_n - S_q)^2 \, dx$$

$$= \sum_{q+1}^n a_m^2 \int_{(-\ell,\ell)^p} A_{11} \nabla_{X_1} v_m \nabla_{X_1} v_m + \lambda_m v_m^2 \, dX_1.$$

If we denote by $\|S_n - S_q\|$ the first integral above and if we take $1 + v_m$ as the test function in (3.57) we derive

$$\|S_n - S_q\| = \sum_{q+1}^n a_m^2 \int_{(-\ell,\ell)^p} -\lambda_m v_m \, dX_1$$

$$\le C \sum_{q+1}^n a_m^2 \lambda_m \le \varepsilon \quad \text{for } q, n \text{ large enough.} \tag{3.70}$$

(This is due to (3.59), the convergence of this series taking place in $H^1(\omega)$.) We have thus established (3.62). So, taking into account (3.62) and (3.69) we have

$$\nabla_{X_2}(u_\ell - u_\infty) = \sum_{m=1}^{+\infty} a_m v_m \nabla_{X_2} \varphi_m$$

and

$$\int_{\Omega_{\ell_0}} A_{22} \nabla_{X_2}(u_\ell - u_\infty) \nabla_{X_2}(u_\ell - u_\infty) + a(u_\ell - u_\infty)^2 \, dx$$

$$= \sum_{m=1}^{+\infty} a_m^2 \lambda_m \int_{(-\ell_0,\ell_0)^p} v_m^2 \, dX_1$$

$$\leq \int_{(-\ell_0,\ell_0)^p} v_1^2 \, dX_1 \sum_{m=1}^{+\infty} a_m^2 \lambda_m \quad \text{(see Lemma 3.6)}$$

$$= \int_{(-\ell_0,\ell_0)^p} v_1^2 \, dX_1 \int_\omega A_{22} \nabla_{X_2} u_\infty \nabla_{X_2} u_\infty + a u_\infty^2 \, dX_2$$

$$\leq C |u_\infty|_{a,\omega}^2 \int_{(-\ell_0,\ell_0)^p} v_1^2 \, dX_1.$$

Moreover – see Lemma 3.3 – we derive easily from above that

$$|u_\ell - u_\infty|_{2,\Omega_{\ell_0}}^2 \leq C |u_\infty|_{a,\omega} \int_{(-\ell_0,\ell_0)^p} v_1^2 \, dx. \tag{3.71}$$

We estimate this last integral. For that we will rely on the maximum principle to compare v_1 to a known function – more precisely to

$$u_1 = -\sum_{i=1}^{p} \frac{ch(\alpha x_i)}{ch(\alpha \ell)} \tag{3.72}$$

for α suitable.

Indeed we have

LEMMA 3.7. *For α chosen small enough it holds that*

$$0 \geq v_1 \geq u_1. \tag{3.73}$$

PROOF. Clearly u_1 is smooth and less than or equal to -1 on the boundary of $(-\ell, \ell)^p$. From (3.57) we have

$$\int_{(-\ell,\ell)^p} A_{11} \nabla_{X_1} v_1 \nabla_{X_1} v + \lambda_1 v_1 v \, dX_1 = 0 \quad \forall v \in H_0^1((-\ell, \ell)^p).$$

Taking $v = (u_1 - v_1)^+$ we derive

$$\int_{(\ell,\ell)^p} A_{11} \nabla_{X_1}(u_1 - v_1) \nabla_{X_1}(u_1 - v_1)^+ + \lambda_1(u_1 - v_1)(u_1 - v_1)^+ \, dX_1$$

$$= \int_{(-\ell,\ell)^p} A_{11} \nabla_{X_1} u_1 \nabla_{X_1}(u_1 - v_1)^+ + \lambda_1 u_1(u_1 - v_1)^+ \, dX_1$$

$$= \int_{(-\ell,\ell)^p} \{-\nabla_{X_1} \cdot (A_{11} \nabla_{X_1} u_1) + \lambda_1 u_1\}(u_1 - v_1)^+ \, dX_1$$

$$= \int_{(-\ell,\ell)^p} \{-\partial_{x_i}(a_{ij}(X_1) \partial_{x_j} u_1) + \lambda_1 u_1\}(u_1 - v_1)^+ \, dX_1,$$

where the summation in i, j runs from 1 to p. Thus, we get

$$
\int_{(-\ell,\ell)^p} A_{11} \nabla_{X_1}(u_1 - v_1)^+ \nabla_{X_1}(u_1 - v_1)^+ + \lambda_1(u_1 - v_1)^{+2} \, dX_1
$$

$$
= \int_{(-\ell,\ell)^p} \{-a_{ij}(X_1)\partial_{x_i}\partial_{x_j}u_1 - \partial_{x_i}(a_{ij}(X_1))\partial_{x_j}u_1
$$

$$
+ \lambda_1 u_1\}(u_1 - v_1)^+ \, dX_1.
$$

(3.74)

From (3.72) we derive

$$
\partial_{x_j}u_1 = -\alpha \frac{sh(\alpha x_j)}{ch(\alpha \ell)},
$$

$$
\partial_{x_i}\partial_{x_j}u_1 = 0 \qquad i \neq j,
$$

$$
= -\alpha^2 \frac{ch(\alpha x_i)}{ch(\alpha \ell)} \qquad i = j.
$$

Thus from (3.74) we deduce that

$$
\int_{(-\ell,\ell)^p} A_{11} \nabla_{X_1}(u_1 - v_1)^+ \nabla_{X_1}(u_1 - v_1)^+ + \lambda(u_1 - v_1)^{+2} \, dX_1
$$

$$
= \int_{(-\ell,\ell)^p} \left\{ \alpha^2 \sum_{i=1}^{p} a_{ii}(X_1) \frac{ch(\alpha x_i)}{ch(\alpha \ell)} \frac{1}{u_1} \right.
$$

$$
\left. + \alpha \sum_{i,j=1}^{p} \partial_{x_i}(a_{ij}(X_1)) \frac{sh(\alpha x_j)}{ch(\alpha \ell)} \frac{1}{u_1} + \lambda_1 \right\} u_1(u_1 - v_1)^+ \, dX_1.
$$

Since

$$
|\frac{ch(\alpha x_i)}{ch(\alpha \ell)} \frac{1}{u_1}| \leq 1 \qquad , \qquad |\frac{sh(\alpha x_i)}{ch(\alpha \ell)} \frac{1}{u_1}| \leq 1
$$

it is clear that for α small enough it holds – see (3.52)

$$
\left\{ \alpha^2 \sum_{i=1}^{p} a_{ii}(X_1) \frac{ch(\alpha x_i)}{ch(\alpha \ell)} \frac{1}{u_1} + \alpha \sum_{i,j=1}^{p} \partial_{x_i}(a_{ij}(X_1)) \frac{sh(\alpha x_j)}{ch(\alpha \ell)} \frac{1}{u_1} + \lambda_1 \right\} > 0
$$

on $(-\ell, \ell)^p$. It follows that – note that $u_1 < 0$

$$
\int_{(-\ell,\ell)^p} A_{11} \nabla_{X_1}(u_1 - v_1)^+ \nabla_{X_1}(u_1 - v_1)^+ + \lambda_1(u_1 - v_1)^{+2} \, dX_1 \leq 0
$$

$$
\Rightarrow \qquad 0 \geq v_1 \geq u_1 \quad \text{on} \quad (-\ell, \ell)^p.
$$

This completes the proof of Lemma 3.7.   □

END OF THE PROOF OF THEOREM 3.5. We go back to (3.71). We have then

$$
|u_\ell - u_\infty|_{2,\Omega_{\ell_0}}^2 \leq C \int_{(-\ell_0,\ell_0)^p} v_1^2 \, dX_1 \leq C \int_{(-\ell_0,\ell_0)^p} u_1^2 \, dX_1
$$

$$
\leq \frac{C}{(ch(\alpha \ell))^2} \int_{(-\ell_0,\ell_0)^p} (\sum_{i=1}^{p} ch(\alpha x_i))^2 \, dX_1 \leq C(\ell_0)e^{-2\alpha \ell}.
$$

(3.75)

This shows the convergence of u_ℓ toward u_∞ with an exponential rate for the $L^2(\Omega_{\ell_0})$-norm. To get the exponential rate for the $H^1(\Omega_{\ell_0})$-norm we use the equation. Indeed let us denote by η a smooth function of x_1 such that

$$0 \leq \eta \leq 1, \qquad \eta = 1 \quad \text{on} \quad \Omega_{\ell_0}, \qquad \eta = 0 \quad \text{in} \quad \Omega_\ell \setminus \Omega_{\ell_0+1}.$$

Then, from (3.61) we derive

$$\begin{aligned}
\int_{\Omega_\ell} & A_{11} \nabla_{\mathrm{X}_1}(u_\ell - u_\infty) \nabla_{\mathrm{X}_1}\{(u_\ell - u_\infty)\eta^2\} \\
& + A_{22} \nabla_{\mathrm{X}_2}(u_\ell - u_\infty) \nabla_{\mathrm{X}_2}(u_\ell - u_\infty)\eta^2 \\
& + a(u_\ell - u_\infty)^2 \eta^2 \, dx = 0.
\end{aligned} \tag{3.76}$$

Performing the derivation in the first term above we obtain

$$\begin{aligned}
\int_{\Omega_{\ell_0+1}} & \{A_{11} \nabla_{\mathrm{X}_1}(u_\ell - u_\infty) \nabla_{\mathrm{X}_1}(u_\ell - u_\infty) \\
& + A_{22} \nabla_{\mathrm{X}_2}(u_\ell - u_\infty) \nabla_{\mathrm{X}_2}(u_\ell - u_\infty) + a(u_\ell - u_\infty)^2\}\eta^2 \, dx \\
= -2 & \int_{\Omega_{\ell_0+1}} A_{11} \nabla_{\mathrm{X}_1}(u_\ell - u_\infty) \nabla_{\mathrm{X}_1}\eta(u_\ell - u_\infty)\eta \\
\leq C & \int_{\Omega_{\ell_0+1}} |\nabla_{\mathrm{X}_1}(u_\ell - u_\infty)|\eta|u_\ell - u_\infty| \, dx
\end{aligned}$$

for some constant C – see (3.3).

Using the ellipticity condition of the matrix A we derive easily

$$\begin{aligned}
\int_{\Omega_{\ell_0+1}} & \{\lambda|\nabla(u_\ell - u_\infty)|^2 + a(u_\ell - u_\infty)^2\}\eta^2 \, dx \\
& \leq C \int_{\Omega_{\ell_0+1}} |\nabla(u_\ell - u_\infty)|\eta|u_\ell - u_\infty| \, dx \\
& \leq C \left\{ \int_{\Omega_{\ell_0+1}} |\nabla(u_\ell - u_\infty)|^2\eta^2 \, dx \right\}^{1/2} |u_\ell - u_\infty|_{2,\Omega_{\ell_0+1}}
\end{aligned}$$

by the Cauchy–Schwarz inequality. It follows that

$$\min(\lambda, 1) \int_{\Omega_{\ell_0+1}} \{|\nabla(u_\ell - u_\infty)|^2 + a(u_\ell - u_\infty)^2\}\eta^2 \, dx \leq C^2 |u_\ell - u_\infty|_{2,\Omega_{\ell_0+1}}^2.$$

Hence finally for some constant C independent of ℓ we have – since $\eta = 1$ on Ω_{ℓ_0}

$$|(u_\ell - u_\infty)|_{a,\Omega_{\ell_0}}^2 \leq C|u_\ell - u_\infty|_{2,\Omega_{\ell_0+1}}^2 \leq Ce^{-2\alpha\ell}.$$

This completes the proof of the theorem.   □

REMARK 3.4. In the above proof we estimated the norm in $H^1(\Omega_{\ell_0})$ in terms of the $L^2(\Omega_{\ell_0+1})$-norm – i.e., in terms of the L^2-norm of a larger domain. These kind of estimates are very common and useful in many aspects of the theory of elliptic equations.

3.3. Convergence in higher Sobolev spaces

A mathematically fruitful method is to argue formally. It gives some inspiration that can be justified later on relatively easily. Let us rely on this principle to see how things are working (see also Theorem 2.12). Suppose that u_ℓ is the solution to (3.23) and u_∞ the solution to (3.11). We would like then to estimate the second derivatives of $u_\ell - u_\infty$. For that we start from an equation that we can derive from (3.23), (3.11), namely

$$-\sum_{i,j=1}^{n} \partial_{x_i}(a_{ij}(x)\partial_{x_j}u_\ell) + au_\ell = -\sum_{i,j=p+1}^{n} \partial_{x_i}(a_{ij}(x_2)\partial_{x_j}u_\infty) + au_\infty. \qquad (3.77)$$

Since u_∞ is independent of x_1 we get in the weak sense

$$-\sum_{i,j=1}^{n} \partial_{x_i}(a_{ij}(x)\partial_{x_j}(u_\ell - u_\infty)) + a(u_\ell - u_\infty) = 0 \quad \text{in } \Omega_\ell. \qquad (3.78)$$

Let us differentiate this equation in the direction x_k. We obtain, setting $w = u_\ell - u_\infty$,

$$-\partial_{x_i}(a_{ij}\partial_{x_j}(\partial_{x_k}w)) + a\partial_{x_k}w = \partial_{x_i}((\partial_{x_k}a_{ij})\partial_{x_j}w) - (\partial_{x_k}a)w \qquad (3.79)$$

(we made the summation convention and assumed a_{ij}, a, differentiable). Let us then denote by η a smooth function with compact support in Ω_{ℓ_0}. Using as test function in the weak formulation of (3.79) the function

$$(\partial_{x_k}w)\eta^2 \qquad (3.80)$$

we obtain

$$\int_{\Omega_{\ell_0}} A\nabla w_k \nabla(w_k\eta^2) + aw_k^2\eta^2 = \int_{\Omega_{\ell_0}} -A_k\nabla w\nabla(w_k\eta^2) - (\partial_{x_k}a)ww_k\eta^2 \qquad (3.81)$$

where we have set $w_k = \partial_{x_k}w$, $A = (a_{ij})$, $A_k = (\partial_{x_k}a_{ij})$.

We remark that

$$\nabla(w_k\eta^2) = \eta\nabla(w_k\eta) + w_k\eta\nabla\eta$$

thus we derive

$$\int_{\Omega_{\ell_0}} A\nabla w_k\nabla(w_k\eta)\eta + A\nabla w_k\nabla\eta w_k\eta + aw_k^2\eta^2\,dx$$

$$= -\int_{\Omega_{\ell_0}} A_k\nabla w\nabla(w_k\eta)\eta + A_k\nabla w\nabla\eta w_k\eta + (\partial_{x_k}a)ww_k\eta^2\,dx.$$

Finally we use

$$\eta\nabla w_k = \nabla(w_k\eta) - w_k\nabla\eta$$

to get

$$\int_{\Omega_{\ell_0}} A\nabla(w_k\eta)\nabla(w_k\eta) + aw_k^2\eta^2\,dx$$

$$= \int_{\Omega_{\ell_0}} A\nabla\eta\nabla(w_k\eta)w_k\,dx - \int_{\Omega_{\ell_0}} A\nabla(w_k\eta)\nabla\eta w_k\,dx$$

$$- \int_{\Omega_{\ell_0}} (\partial_{x_k}a)ww_k\eta^2\,dx + \int_{\Omega_{\ell_0}} A\nabla\eta\nabla\eta w_k^2\,dx \qquad (3.82)$$

$$- \int_{\Omega_{\ell_0}} A_k\nabla w\nabla(w_k\eta)\eta\,dx - \int_{\Omega_{\ell_0}} A_k\nabla w\nabla\eta w_k\eta\,dx.$$

Recalling the definition of w_k, the ellipticity of A and assuming the functions a_{ij}, $\partial_{x_k}a_{ij}$, a, $\partial_{x_k}a$ bounded we obtain

$$|w_k\eta|_{a,\Omega_{\ell_0}}^2 \le C|w_k\eta|_{a,\Omega_{\ell_0}}|w|_{a,\Omega_{\ell_0}} + C|w|_{a,\Omega_{\ell_0}}^2 \qquad (3.83)$$

(C is some constant depending on η). By the Young inequality

$$ab \le \varepsilon a^2 + \frac{b^2}{\varepsilon} \qquad (3.84)$$

we derive

$$|w_k\eta|_{a,\Omega_{\ell_0}}^2 \le C\varepsilon|w_k\eta|_{a,\Omega_{\ell_0}}^2 + C_\varepsilon|w|_{a,\Omega_{\ell_0}}^2. \qquad (3.85)$$

C_ε is some constant depending on ε. Choosing $C\varepsilon = \frac{1}{2}$ we get

$$|w_k\eta|_{a,\Omega_{\ell_0}}^2 \le C|w|_{a,\Omega_{\ell_0}}^2 \le \frac{C}{\ell^{2r}} \qquad (3.86)$$

(see Theorem 3.2). Thus choosing $\eta = 1$ on $\Omega'_{\ell_0} \Subset \Omega_{\ell_0}$ we obtain

$$|u_\ell - u_\infty|_{H^2(\Omega'_{\ell_0})} \le \frac{C}{\ell^r}$$

for any subcompact domain of Ω_{ℓ_0}. So, following the different steps of these formal estimates we should be able to prove

THEOREM 3.8. *Under the assumptions of Theorem 3.2, let u_ℓ, u_∞ be the solutions to (3.23) and (3.11) respectively. Assume that a_{ij}, a are Lipschitz continuous in Ω in such a way that for any $k = 1, \ldots, n$ it holds that*

$$|\partial_{x_k}a_{ij}|_\infty, |\partial_{x_k}a|_\infty \le D \quad \forall i,j = 1,\ldots,n$$

where D is some positive constant. Then, for any $r > 0$, $\ell_0 > 0$, $\Omega'_{\ell_0} \Subset \Omega_{\ell_0}$, $u_\ell - u_\infty \in H^2(\Omega'_{\ell_0})$ and there exists a constant C independent of ℓ such that

$$|u_\ell - u_\infty|_{H^2(\Omega'_{\ell_0})} \le \frac{C}{\ell^r}. \qquad (3.87)$$

PROOF. We mimic the proof given above. For that we are going to replace the derivative in the x_k direction by finite differences. In other words for any function v we define for $h > 0$, h small

$$\tau_h v(x) = v(x + he_k) \qquad (3.88)$$

where e_k denotes the vector of components $(0,\ldots,1,0\ldots0)$ the 1 being the k-th coordinate of the vector. Moreover, we set

$$\delta_{x_k}(v) = \delta_{x_k}^h(v) = \frac{T_h v(x) - v(x)}{h}. \tag{3.89}$$

We first look for the analogue of (3.79). Let us start by a lemma:

LEMMA 3.9 (Discrete integration by parts). *Let $\mathcal{O}$ be an open subset of $\mathbb{R}^n$, $f \in L^2(\mathcal{O})$, $\varphi \in \mathcal{D}(\mathcal{O})$. Then, for h small enough it holds that*

$$\int_{\mathcal{O}} f\delta_{x_k}^h(\varphi)\,dx = -\int_{\mathcal{O}} \delta_{x_k}^{-h}(f)\varphi\,dx. \tag{3.90}$$

PROOF. We choose $h < \operatorname{dist}(\operatorname{Supp}\varphi, \partial\mathcal{O})$ where $\operatorname{Supp}\varphi$ denotes the support of φ and $\partial\mathcal{O}$ the boundary of $\mathcal{O}$. Then, $\varphi(x + he_k), \varphi(x)$ have their supports in $\mathcal{O}$ and it holds that

$$\int_{\mathcal{O}} f\delta_{x_k}(\varphi)\,dx = \int_{\mathcal{O}} f(x)\frac{\varphi(x + he_k) - \varphi(x)}{h}\,dx$$
$$= \frac{1}{h}\left\{\int_{\mathcal{O}} f(x)\varphi(x + he_k)\,dx - \int_{\mathcal{O}} f(x)\varphi(x)\,dx\right\}.$$

Changing $x + he_k$ in x in the first integral leads to

$$\int_{\mathcal{O}} f\delta_{x_k}(\varphi)\,dx = \frac{1}{h}\left\{\int_{\mathcal{O}} f(x - he_k)\varphi(x)\,dx - \int_{\mathcal{O}} f(x)\varphi(x)\,dx\right\}$$
$$= -\int_{\mathcal{O}} \delta_{x_k}^{-h}(f)\varphi(x)\,dx.$$

This completes the proof of the Lemma. $\qquad\square$

We then go back to the proof of Theorem 3.8. From (3.23), (3.11) we deduce that for any $v \in V_{\Omega_\ell}$ it holds that – if $w = u_\ell - u_\infty$

$$\int_{\Omega_\ell} a_{ij}\partial_{x_j}w\partial_{x_i}v + awv = 0 \quad \forall v \in V_{\Omega_\ell}.$$

In particular if $v \in H_0^1(\Omega_\ell)$ and has compact support – for h small enough we have

$$\int_{\Omega_\ell} a_{ij}\partial_{x_j}w\partial_{x_i}(-\delta_{x_k}^{-h}v) + aw(-\delta_{x_k}^{-h}v)\,dx = 0. \tag{3.91}$$

Using Lemma 3.9 we obtain easily (note that ∂_{x_i} and $\delta_{x_k}^{-h}$ commute)

$$\int_{\Omega_\ell} \delta_{x_k}^h(a_{ij}\partial_{x_j}w)\partial_{x_i}v + \delta_{x_k}^h(aw)v\,dx = 0 \tag{3.92}$$

for any v with compact support in Ω_ℓ.

REMARK 3.5. If the direction x_k is such that $k = 1,\ldots,p$, then (3.92) holds for any $v \in V_{\Omega_{\ell_1}}$ with $\ell_1 < \ell$ and any $h < \ell - \ell_1$.

It is easy to see that for two functions f, g it holds that

$$\delta^h_{x_k}(fg) = f(x + he_k)\delta^h_{x_k}(g) + \delta^h_{x_k}f(x)g(x). \tag{3.93}$$

Thus from (3.93) we derive

$$\int_{\Omega_\ell} \tau_h(a_{ij})\partial_{x_j}(\delta_{x_k}w)\partial_{x_i}v + \tau_h(a)\delta_{x_k}wv\,dx$$
$$= -\int_{\Omega_\ell} \delta_{x_k}(a_{ij})\partial_{x_j}w\partial_{x_i}v + \delta_{x_k}(a)wv\,dx \tag{3.94}$$

for any v with compact support S in Ω_ℓ and any $h < \text{dist}(S, \partial\Omega_\ell)$. This is exactly the discrete analogue to (3.79). Then, we consider a compact subset $\Omega'_{\ell_0} \Subset \Omega_{\ell_0}$ and η a smooth function with compact support in Ω_{ℓ_0} such that

$$0 \leq \eta \leq 1, \qquad \eta = 1 \quad \text{on} \quad \Omega'_{\ell_0}, \qquad |\nabla\eta| \leq C$$

where C is some constant. We consider then in (3.94)

$$v = \delta_{x_k}(w)\eta^2. \tag{3.95}$$

Clearly, for $h < 2\,\text{dist}(\text{Supp}\,\eta, \partial\Omega_{\ell_0})$, v is a test function for (3.94) and we obtain, setting

$$\tau A = (\tau_h a_{ij}), \qquad w_k = \delta_{x_k}w, \qquad A_k = (\delta_{x_k}a_{ij}), \qquad \tau a = \tau_h a,$$

$$\int_{\Omega_{\ell_0}} \tau A\nabla w_k\nabla(w_k\eta^2) + \tau aw_k^2\eta^2\,dx = -\int_{\Omega_{\ell_0}} A_k\nabla w\nabla(w_k\eta^2) + (\delta_{x_k}a)ww_k\eta^2\,dx.$$

This is exactly (3.81). Thus performing the manipulations below (3.81) we arrive at the analogue of (3.82) which is

$$\int_{\Omega_{\ell_0}} \tau A\nabla(w_k\eta)\nabla(w_k\eta) + \tau aw_k^2\eta^2\,dx$$
$$= \int_{\Omega_{\ell_0}} \tau A\nabla\eta\nabla(w_k\eta)w_k\,dx - \int_{\Omega_{\ell_0}} A\nabla(w_k\eta)\nabla\eta w_k\,dx$$
$$- \int_{\Omega_{\ell_0}} (\delta_{x_k}a)ww_k\eta^2\,dx + \int_{\Omega_{\ell_0}} \tau A\nabla\eta\nabla\eta w_k^2\,dx \tag{3.96}$$
$$- \int_{\Omega_{\ell_0}} A_k\nabla w\nabla(w_k\eta)\eta\,dx - \int_{\Omega_{\ell_0}} A_k\nabla w\nabla\eta(w_k\eta)\,dx.$$

Using the ellipticity conditions and writing $\tau a = a + \tau a - a$ we obtain easily

$$|w_k\eta|^2_{a,\Omega_{\ell_0}} \leq C\{|w_k\eta|_{a,\Omega_{\ell_0}}|w_k|_{2,\eta} + |w|_{a,\Omega_{\ell_0}}|w_k|_{2,\eta}$$
$$+ |w_k|^2_{2,\eta} + |w_k\eta|_{a,\Omega_{\ell_0}}|w|_{a,\Omega_{\ell_0}}\}$$

where $|w_k|^2_{2,\eta} = \int_{\text{Supp}\,\eta} w_k^2(x)\,dx$. Using the Young inequality (3.84) we arrive easily at

$$|w_k\eta|^2_{a,\Omega_{\ell_0}} \leq C\{|w_k|^2_{2,\eta} + |w|^2_{a,\Omega_{\ell_0}}\}. \tag{3.97}$$

Let us show

LEMMA 3.10. *Let $\mathcal{O}' \Subset \mathcal{O}$ be two bounded open sets of $\mathbb{R}^n$. For $h < \text{dist}(\mathcal{O}', \partial\mathcal{O})$, $w \in H^1(\mathcal{O})$ it holds that:*

$$|w_k|_{2,\mathcal{O}'} = |\delta_{x_k} w|_{2,\mathcal{O}'} \leq |\partial_{x_k} w|_{2,\mathcal{O}}. \tag{3.98}$$

PROOF. For $x \in \mathcal{O}'$, $x + he_k \in \mathcal{O}$ and for w smooth it holds that

$$\delta_{x_k} w(x) = \frac{1}{h} \int_0^1 \frac{d}{dt} w(x + the_k)\,dt = \int_0^1 \frac{\partial w}{\partial x_k}(x + the_k)\,dt.$$

Squaring and using the Cauchy–Schwarz inequality we obtain

$$(\delta_{x_k} w(x))^2 \leq \int_0^1 (\partial_{x_k} w)^2(x + the_k)\,dt.$$

Integrating in x and using Fubini's theorem we obtain

$$|w_k|_{2,\mathcal{O}'} \leq \int_0^1 \int_{\mathcal{O}} (\partial_{x_k} w)^2(x)\,dx\,dt \leq |\partial_{x_k} w|_{2,\mathcal{O}}^2.$$

We complete the proof by using the density of the smooth functions in $H^1(\mathcal{O})$. $\square$

END OF THE PROOF OF THEOREM 3.8. Going back to (3.97) we derive from the lemma for h small

$$|w_k \eta|_{a,\Omega_{\ell_0}}^2 \leq C|w|_{a,\Omega_{\ell_0}}^2 \leq \frac{C}{\ell^{2r}}.$$

In particular

$$\|\nabla w_k\|_{2,\Omega_{\ell_0}'}^2 \leq \frac{C}{\ell^{2r}}.$$

In other words the functions $\partial_{x_i} w_k$ are bounded in $L^2(\Omega_{\ell_0}')$, thus up to a subsequence, they converge in $L^2(\Omega_{\ell_0}')$-weak toward some function f_{ik}. Moreover passing to the lim inf in

$$|\partial_{x_i} w_k|_{2,\Omega_{\ell_0}'}^2 \leq \frac{C}{\ell^{2r}}$$

we obtain

$$|f_{ik}|_{2,\Omega_{\ell_0}'}^2 \leq \frac{C}{\ell^{2r}}. \tag{3.99}$$

(In a Hilbert space the norm is continuous and thus weakly lower semi-continuous.) It is easy to verify that

$$\partial_{x_i} w_k \to \partial_{x_i} \partial_{x_k} w \quad \text{in } \mathcal{D}'(\Omega_{\ell_0}').$$

Thus, by uniqueness of the limit

$$\partial_{x_i} \partial_{x_k} w = f_{ik} \in L^2(\Omega_{\ell_0}').$$

We have thus established that $w = u_\ell - u_\infty \in H^2(\Omega_{\ell_0}')$ and by (3.99) the proof of the theorem is complete. $\square$

REMARK 3.6. By Remark 3.5, using $\eta = \eta(x_1)$ with

$$0 \le \eta \le 1, \qquad \eta = 1 \text{ on } \Omega_{\ell_0}, \qquad \eta = 0 \text{ outside } \Omega_{\ell_0+1},$$

one would get easily, under the assumptions of Theorem 3.8,

$$\partial_{x_i}\partial_{x_k}(u_\ell - u_\infty) \in L^2(\Omega_{\ell_0}) \quad \forall\, i = 1,\ldots,n \quad \forall\, k = 1,\ldots,p$$

and an estimate

$$|\partial_{x_i}\partial_{x_k}(u_\ell - u_\infty)|_{2,\Omega_{\ell_0}} \le \frac{C}{\ell^r}, \tag{3.100}$$

valid for $i = 1,\ldots,n$, $k = 1,\ldots,p$.

REMARK 3.7. It is clear that assuming more regularity on the a_{ij}, a one can use this technique to get higher estimates. In other words it holds that

$$|u_\ell - u_\infty|_{H^k(\Omega'_{\ell_0})} \le \frac{C}{\ell^r}$$

for any k such that the $k-1$ derivatives of a_{ij}, a are bounded. Choosing $k > \frac{n}{2}$ due to the Sobolev embedding theorem we get that

$$|u_\ell - u_\infty|_{\infty,\Omega'_{\ell_0}} \le \frac{C}{\ell^r}$$

i.e., u_ℓ converges toward u_∞ uniformly locally.

Let us now consider the particular case where $V_{\Omega_\ell} = H_0^1(\Omega_\ell)$ i.e., the case of homogeneous boundary conditions for u_ℓ and u_∞. Then we can prove

THEOREM 3.11. *Under the assumptions of Theorem 3.8 let u_ℓ, u_∞ be the solution to*

$$u_\ell \in H_0^1(\Omega_\ell), \quad \int_{\Omega_\ell} a_{ij}(x)\partial_{x_j}u_\ell\partial_{x_i}v + au_\ell v\, dx = \langle f, v\rangle \quad \forall\, v \in H_0^1(\Omega_\ell), \tag{3.101}$$

$$u_\infty \in H_0^1(\omega), \quad \int_\omega a_{ij}(x)\partial_{x_j}u_\infty\partial_{x_i}v + au_\infty v\, dx = \langle f, v\rangle \quad \forall\, v \in H_0^1(\omega), \tag{3.102}$$

then if ω is smooth enough for any $\ell_0 > 0$, $r > 0$ it holds that

$$u_\ell - u_\infty \in H^2(\Omega_{\ell_0}) \text{ and } |u_\ell - u_\infty|_{H^2(\Omega_{\ell_0})} \le \frac{C}{\ell^r} \tag{3.103}$$

(in (3.102) the summation runs between $p+1$ and n.)

PROOF. From (3.100) we have

$$|\partial_{x_i}\partial_{x_k}(u_\ell - u_\infty)|_{2,\Omega_{\ell_0}} \le \frac{C}{\ell^r} \tag{3.104}$$

for $i = 1,\ldots,n$, $k = 1,\ldots,p$. Thus, due to the symmetry property of the second derivative it is enough to show that

$$|\partial_{x_i}\partial_{x_k}(u_\ell - u_\infty)|_{2,\Omega_{\ell_0}} \le \frac{C}{\ell^r} \quad \forall\, i,k = p+1,\ldots,n. \tag{3.105}$$

For that we remark that $u_\ell - u_\infty$ satisfies in the distributional sense

$$\sum_{i,j=p+1}^{n} \partial_{x_i}(a_{ij}\partial_{x_j}(u_\ell - u_\infty)) =$$

$$- \sum_{(i,j)\notin\{p+1,\ldots,n\}^2} \partial_{x_i}(a_{ij}\partial_{x_j}(u_\ell - u_\infty)) + a(u_\ell - u_\infty).$$

In particular for a.e. $X_1 \in (-\ell_0, \ell_0)^p$, $(u_\ell - u_\infty)(X_1, \cdot)$ is solution of an elliptic problem with right-hand side in $L^2(\omega)$. It follows that for a.e. X_i it holds that

$$|(u_\ell - u_\infty)(X_1, \cdot)|_{H^2(\omega)}$$

$$\leq C\left\{|(u_\ell - u_\infty)(X_1, \cdot)|_\omega^2 + \sum_{(i,j)\notin\{p+1,\ldots,n\}^2} |\partial_{x_i}\partial_{x_j}(u_\ell - u_\infty)(X_1, \cdot)|_\omega^2\right\}$$

and by integration in $X_1 \in (-\ell_0, \ell_0)^p$, (3.105) holds thanks to (3.104). This completes the proof of Theorem 3.11.   □

REMARK 3.8. As in Remark 3.7 we can also bootstrap in this case to obtain an estimate in $H^k(\Omega_{\ell_0})$. The details are left to the reader.

Open problems

1. Can we further extend the exponential rate of convergence of u_ℓ toward u_∞ – i.e., under more general conditions than the ones of Theorem 3.5?

2. We consider u_ℓ the solution to

$$\begin{cases} \int_{\Omega_\ell} A^\ell \nabla u_\ell \nabla v + a^\ell u_\ell v \, dx = \langle f^\ell, v \rangle \quad \forall v \in V_{\Omega_\ell}, \\ u_\ell \in V_{\Omega_\ell}. \end{cases}$$

Trace the minimal convergence assumptions

$$A_{22}^\ell \to A_{22}(X_2), \qquad a^\ell \to a(X_2), \qquad f^\ell \to f(X_2)$$

in order for u_ℓ to convergence toward u_∞ the solution to

$$\begin{cases} \int_\omega A_{22}\nabla_{X_2} u_\infty \nabla_{X_2} v + a u_\infty v \, dx = \langle f, v \rangle \quad \forall v \in V, \\ u_\infty \in V. \end{cases}$$

Chapter 4

Nonlinear Elliptic Problems

In this chapter we will introduce various nonlinear problems and analyze the existence and uniqueness issues. Among them are the variational inequalities introduced in [35]. Let us begin with the abstract theory of the variational inequalities.

4.1. Variational inequalities

4.1.1. A generalisation of the Lax–Milgram theorem. In this section we suppose that $a(u, v)$ is a bilinear form on some Hilbert space H satisfying (1.1), (1.2) – i.e., we suppose that we are under the assumptions of Theorem 1.1. Then we have:

THEOREM 4.1 (Variational inequality). *Suppose that we are under the assumptions of Theorem 1.1. Let $K \neq \emptyset$ be a closed convex subset of H. Then, for every $f \in H'$, there exists a unique u solution to*

$$\begin{cases} a(u, v - u) \geq \langle f, v - u \rangle & \forall v \in K, \\ u \in K. \end{cases} \tag{4.1}$$

Moreover, if a is symmetric then u is the unique minimizer of

$$J(v) = \frac{1}{2} a(v, v) - \langle f, v \rangle \tag{4.2}$$

on K.

PROOF. Recall that a convex set is a set such that

$$u, v \in K \quad \Rightarrow \quad \alpha u + (1 - \alpha)v \in K \quad \forall \alpha \in [0, 1]. \tag{4.3}$$

Suppose first that

$$a(u, v) = (u, v) \tag{4.4}$$

where $(\cdot, \cdot)$ denotes the scalar product on H. Due to the Riesz representation theorem (see [6]) there exists a unique $\tilde{f} \in H$ such that

$$(\tilde{f}, v) = \langle f, v \rangle \quad \forall v \in H. \tag{4.5}$$

Then, in the case of (4.4), (4.1) can be written

$$\begin{cases} (u - \tilde{f}, v - u) \geq 0 & \forall v \in K, \\ u \in K, \end{cases} \tag{4.6}$$

and this problem admits a unique solution given by

$$u = P_K(\tilde{f}) \tag{4.7}$$

the projection of $\tilde{f}$ on K. With this in mind, for $u \in H$, we introduce

$$w = S(u) \qquad (4.8)$$

the solution to

$$\begin{cases} (w, v - w) \geq (u, v - w) - \varrho\{a(u, v - w) - \langle f, v - w \rangle\}, \\ w \in K, \end{cases} \qquad (4.9)$$

where ϱ is some positive number that we will fix later on. Since

$$v \mapsto (u, v) - \varrho\{a(u, v) - \langle f, v \rangle\}$$

is a continuous linear form on H, it is clear that (4.9) admits a unique solution w for any $u \in H$. If for some $\varrho > 0$ we show that $u \mapsto S(u) = w$ admits a fixed point, this fixed point will be clearly a solution to (4.1). For that – by the Banach fixed point theorem – it is enough to show that S is a contraction. So, consider $u_1, u_2 \in H$ and $w_1 = S(u_1)$, $w_2 = S(u_2)$. From (4.9) we derive

$$w_1 \in K, \quad (w_1, v - w_1) \geq (u_1, v - w_1) - \varrho\{a(u_1, v - w_1) - \langle f, v - w_1 \rangle\} \quad \forall v \in K,$$

$$w_2 \in K, \quad (w_2, v - w_2) \geq (u_2, v - w_2) - \varrho\{a(u_2, v - w_2) - \langle f, v - w_2 \rangle\} \quad \forall v \in K.$$

Taking $v = w_2$ in the first inequality, $v = w_1$ in the second and adding together the two inequalities obtained we get

$$|w_1 - w_2|^2 \leq (u_1 - u_2, w_1 - w_2) - \varrho a(u_1 - u_2, w_1 - w_2). \qquad (4.10)$$

We consider the bilinear form

$$(u, v) - \varrho a(u, v).$$

By the Riesz representation theorem there exists $f_u \in H$ such that

$$a(u, v) = (f_u, v) \quad \forall v \in H. \qquad (4.11)$$

Moreover by (1.1)

$$|f_u|^2 = (f_u, f_u) = a(u, f_u) \leq \Lambda|u|\,|f_u|$$

and thus

$$|f_u| \leq \Lambda|u|.$$

From (4.11) we derive

$$|(u, v) - \varrho a(u, v)| = |(u - \varrho f_u, v)| \leq |u - \varrho f_u|\,|v|.$$

Moreover, by (1.2)

$$|u - \varrho f_u|^2 = |u|^2 - 2\varrho(f_u, u) + \varrho^2|f_u|^2$$
$$\leq |u|^2 - 2\varrho\lambda(u, u) + \varrho^2\Lambda^2|u|^2$$
$$\leq (1 - 2\varrho\lambda + \varrho^2\Lambda^2)|u|^2.$$

It is clear that for $\varrho \in (0, \frac{2\lambda}{\Lambda^2})$ it holds that

$$0 < 1 - 2\varrho\lambda + \varrho^2\Lambda^2 < 1.$$

Denoting this number by ν^2, $(\nu \in (0,1))$ we derive from above that

$$|(u,v) - \varrho a(u,v)| \leq \nu |u|\, |v|$$

and from (4.10) it follows that

$$|w_1 - w_2|^2 \leq \nu |u_1 - u_2|\, |w_1 - w_2| \quad \Rightarrow \quad |w_1 - w_2| \leq \nu |u_1 - u_2|.$$

Since $0 < \nu < 1$, S is a contraction and the existence of u solution to (4.1) follows. To show uniqueness it is enough to notice that if u_1, u_2 are two solutions of (4.1) then it holds that

$$a(u_1, u_2 - u_1) \geq \langle f, u_2 - u_1 \rangle,$$
$$a(u_2, u_1 - u_2) \geq \langle f, u_1 - u_2 \rangle$$

and by adding

$$\lambda |u_1 - u_2|^2 \leq a(u_1 - u_2, u_1 - u_2) \leq 0.$$

Hence $u_1 = u_2$ and the uniqueness follows. Finally rewriting the computation (1.11) for u solution to (4.1) and $v \in K$ we get when a is symmetric

$$
\begin{aligned}
J(v) = J(u + v - u) &= \frac{1}{2} a(u + v - u, u + v - u) - \langle f, v \rangle \\
&= \frac{1}{2} a(u,u) + a(u, v - u) + \frac{1}{2} a(v - u, v - u) - \langle f, v \rangle \\
&\geq \frac{1}{2} a(u,u) + \langle f, v - u \rangle + \frac{1}{2} a(v - u, v - u) - \langle f, v \rangle \\
&= J(u) + \frac{1}{2} a(v - u, v - u) > J(v) \quad \forall v \in K, \quad v \neq u.
\end{aligned}
\tag{4.12}
$$

This completes the proof of the theorem. $\qquad\square$

A huge literature is devoted to variational inequalities. For details the reader can consult [23], [20], [18], [22], [4], [7], [30], [35]. We will restrict ourselves to the theory in H^1.

4.1.2. Some examples of variational inequalities. Let Ω be a Lipschitz open subset of $\mathbb{R}^n$. Under the assumptions of Section 1.3 define $a(u,v)$ by (1.43) that is to say by

$$a(u,v) = \int_\Omega a_{ij}(x)\partial_{x_j} u \partial_{x_i} v + a(x) uv \, dx. \tag{4.13}$$

Clearly $a(u,v)$ is defined for any $u, v \in H^1(\Omega)$.

If we choose – with the notation of Section 1.2

$$V = V_\Omega = H^1_0(\Omega), H^1_0(\Omega; \Gamma_0), H^1(\Omega), W, \tag{4.14}$$

by Theorem 1.8, a is a bilinear, continuous, coercive form on V (of course with the additional assumption $a(x) \not\equiv 0$ in the case where $V = H^1(\Omega)$.) Thus, if K is a closed convex set of V, $K \neq \emptyset$ and if $f \in V'$ the dual of V, then there exists (see Theorem 4.1) a unique u solution to

$$\begin{cases} a(u, v - u) \geq \langle f, v - u \rangle \quad \forall v \in K, \\ u \in K. \end{cases} \tag{4.15}$$

For K we can consider two types of constraints, some local constraints, some nonlocal ones.

- The case of local constraints.

Suppose that for a.e. $x \in \Omega$

$$K(x) \text{ is a closed nonempty convex set of } \mathbb{R}^{n+1}. \tag{4.16}$$

Then we can show

PROPOSITION 4.2. *Let V be defined by (4.14) and for a.e. $x \in \Omega$, let $K(x)$ be such that (4.16) holds. Then, if we set*

$$K = \{ v \in V \mid (v(x), \nabla v(x)) \in K(x) \text{ a.e. } x \in \Omega \} \tag{4.17}$$

K is a closed convex subset of V.

PROOF. The convexity of K is clear. Indeed, if $u, v \in K$, then $u, v \in V$ and thus for any $\alpha \in (0, 1)$

$$\alpha u + (1 - \alpha)v \in V.$$

Moreover it holds that

$$\begin{aligned} ((\alpha u &+ (1 - \alpha)v)(x), \nabla(\alpha u + (1 - \alpha)v)(x)) \\ &= (\alpha u(x) + (1 - \alpha)v(x), \alpha \nabla u(x) + (1 - \alpha)\nabla v(x)) \\ &= \alpha(u(x), \nabla u(x)) + (1 - \alpha)(v(x), \nabla v(x)) \in K(x) \quad \text{a.e. } x \in \Omega. \end{aligned}$$

To show now that K is closed, we can suppose that $K \neq \emptyset$ and consider a sequence such that

$$v_n \in K, \qquad v_n \to v \quad \text{in } V. \tag{4.18}$$

Due to the topology defined on V we have

$$v_n \to v \quad \text{in } H^1(\Omega)$$

and thus, up to a subsequence, we can assume that

$$(v_n(x), \nabla v_n(x)) \to (v, \nabla v(x)) \quad \text{a.e. } x \in \Omega$$

i.e., on $\Omega \setminus N_0$ where N_0 is a set of measure 0.

By assumption we have a.e. $x \in \Omega$

$$(v_n(x), \nabla v_n(x)) \in K(x), \quad K(x) \text{ closed}$$

i.e., for every $x \in \Omega \setminus N_n$, $n = 1, 2, \ldots$, where N_n is a set of measure 0. Then, on $\Omega \setminus \bigcup_{n=0}^{+\infty} N_n$, we have

$$K(x) \ni (v_n(x), \nabla v_n(x)) \to (v(x), \nabla v(x)),$$

and $K(x)$ closed in $\mathbb{R}^{n+1}$. Thus we have

$$(v(x), \nabla v(x)) \in K(x) \quad \forall x \in \Omega \setminus \bigcup_{n=0}^{+\infty} N_n,$$

i.e., $\quad (v(x), \nabla v(x)) \in K(x) \quad$ a.e. $x \in \Omega$

since $\bigcup_{n=0}^{+\infty} N_n$ is a set of measure 0. That is to say $v \in K$ and K is closed. This completes the proof of the proposition. $\qquad\square$

- The case of nonlocal constraints.

Consider a function

$$J : \Omega \times \mathbb{R}^{n+1} \to \mathbb{R}^+ \tag{4.19}$$

such that

$$x \mapsto J(x, q) \quad \text{is measurable } \forall q \in \mathbb{R}^{n+1}, \tag{4.20}$$

$$q \mapsto J(x, q) \quad \text{is convex a.e. } x \in \Omega. \tag{4.21}$$

Then we can show

PROPOSITION 4.3. *Let Ω' be a subset of positive measure in Ω. Set*

$$K = \left\{ v \in V \;\Big|\; \int_{\Omega'} J(x, v(x), \nabla v(x)) \, dx \leq C \right\} \tag{4.22}$$

where C is some positive constant. Then, K is a closed convex set of V.

PROOF. Due to (4.21), $q \to J(x, q)$ is continuous for a.e. $x \in \Omega$ and thus for $v \in V$, $x \mapsto J(x, v(x), \nabla v(x))$ is measurable and (4.22) makes sense. Let $u, v \in K$, then for a.e. $x \in \Omega$ it holds that

$$J(x, \alpha u(x) + (1 - \alpha)v(x), \alpha \nabla u(x) + (1 - \alpha)\nabla v(x))$$
$$\leq \alpha J(x, u(x), \nabla u(x)) + (1 - \alpha)J(x, v(x), \nabla v(x)). \tag{4.23}$$

Since for $u, v \in K$, $J(x, v(x), \nabla v(x))$, $J(x, u(x), \nabla u(x))$ are integrable, so is the left-hand side of (4.23) and it holds that

$$\int_{\Omega'} J(x, \alpha u(x) + (1 - \alpha)v(x), \nabla(\alpha u(x) + (1 - \alpha)v(x)) \, dx$$
$$\leq \alpha \int_{\Omega'} J(x, u(x), \nabla u(x)) \, dx + (1 - \alpha) \int_{\Omega'} J(x, v(x), \nabla v(x)) \, dx$$
$$\leq \alpha C + (1 - \alpha)C = C.$$

This proves the convexity of K.

Next consider a sequence $v_n \in K$ such that

$$v_n \to v \quad \text{in } V. \tag{4.24}$$

We have of course $v \in V$ and moreover

$$\int_{\Omega'} J(x, v_n(x)), \nabla v_n(x)) \, dx \leq C. \tag{4.25}$$

From (4.24) we can deduce that for some subsequence – still labelled n – it holds that

$$(v_n(x), \nabla v_n(x)) \to (v(x), \nabla v(x)) \quad \text{a.e. in } \Omega.$$

Thus, by Fatou's Lemma, we deduce from (4.25)

$$\int_{\Omega'} J(x, v(x), \nabla v(x)) \, dx = \int_{\Omega'} \liminf_{n \to +\infty} J(x, v_n(x), \nabla v_n(x)) \, dx$$

$$\leq \liminf_{n \to +\infty} \int_{\Omega'} J(x, v_n(x), \nabla v_n(x)) \, dx \leq C.$$

This shows that $v \in K$ and thus K is closed. This completes the proof of Proposition 4.3. $\qquad \square$

Let us now consider some classical examples:

1) The obstacle problem

Let $f \in L^2(\Omega)$. Then, for φ a measurable function, for a.e. $x \in \Omega$

$$K(x) = [\varphi(x), +\infty) \times \mathbb{R}^n$$

is a closed convex set of $\mathbb{R}^{n+1}$ and by Theorem 4.1 if

$$K = K_\varphi = \{ v \in H_0^1(\Omega) \mid v(x) \geq \varphi(x) \text{ a.e. } x \in \Omega \} \tag{4.26}$$

is not empty, then there exists a unique u solution to

$$\begin{cases} \displaystyle\int_\Omega \nabla u \cdot \nabla(v - u) \, dx \geq \int_\Omega f \cdot v - u \, dx \quad \forall v \in K_\varphi, \\ u \in K_\varphi. \end{cases} \tag{4.27}$$

The problem (4.27) is called the one-obstacle problem. u represents the vertical displacement of an elastic membrane spanned on a horizontal frame Γ, submitted to a density of forces f and obliged to sit over some obstacle represented by the function φ. Ω is the position of the membrane in its undeformed state. Similarly

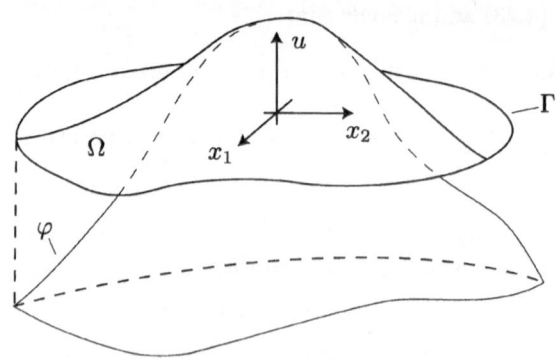

FIGURE 4.1

if ψ is another measurable function we can define for a.e. $x \in \Omega$

$$K(x) = [\varphi(x), \psi(x)] \times \mathbb{R}^n.$$

If K, the closed convex set defined by

$$K = K_\varphi^\psi = \{\, v \in H_0^1(\Omega) \mid \varphi(x) \le v(x) \le \psi(x) \text{ a.e. } x \in \Omega \,\} \qquad (4.28)$$

is not empty, then there exists a unique u solution to

$$\begin{cases} \displaystyle\int_\Omega \nabla u \cdot \nabla(x-u)\, dx \ge \int_\Omega f(v-u)\, dx \quad \forall\, v \in K_\varphi^\psi, \\ u \in K_\varphi^\psi. \end{cases} \qquad (4.29)$$

u is the so-called solution of the double-obstacle problem.

2) Constraints on the gradient

The most famous problem under this header is the elastic-plastic torsion problem. For $f \in L^2(\Omega)$, $a_{ij} = \delta_{ij}$, $a \equiv 0$, $V = H_0^1(\Omega)$ for every $x \in \Omega$ we choose

$$K(x) = \mathbb{R} \times B_1$$

where B_1 denotes the unit ball of $\mathbb{R}^n$ – i.e., if $|\cdot|$ denotes the euclidean norm in $\mathbb{R}^n$

$$B_1 = \{\, y \in \mathbb{R}^n \mid |y| \le 1 \,\}.$$

Then, under these assumptions, there exists a unique u solution to

$$\begin{cases} u \in K = \{\, v \in H_0^1(\Omega) \mid |\nabla v(x)| \le 1 \text{ a.e. } x \in \Omega \,\}, \\ \displaystyle\int_\Omega \nabla u \cdot \nabla(v-u)\, dx \ge \int_\Omega f(v-u)\, dx \quad \forall\, v \in K. \end{cases} \qquad (4.30)$$

When $f = \mathrm{cst}$, u is the horizontal displacement of a bar of section Ω under torsion. f is the angle of torsion. The constraint

$$|\nabla v(x)| \le 1$$

rules the distinction between the elastic behaviour ($|\nabla v(x)| < 1$) and the plastic one ($|\nabla v(x)| = 1$).

3) A nonlocal example

Suppose that we are under the hypotheses of the previous examples. Then, if we set

$$J(x,q) = \left\{ \sum_{i=2}^{n+1} q_i^2 \right\}^{1/2} \quad \forall\, x \in \Omega, \quad \forall\, q = (q_1, \dots, q_{n+1}) \in \mathbb{R}^{n+1},$$

there exists a unique u solution to

$$\begin{cases} u \in K = \left\{\, v \in H_0^1(\Omega) \mid \displaystyle\int_{\Omega'} |\nabla v(x)|^2\, dx \le C \,\right\}, \\ \displaystyle\int_\Omega \nabla u \cdot \nabla(v-u)\, dx \ge \int_\Omega f(v-u)\, dx \quad \forall\, v \in K. \end{cases} \qquad (4.31)$$

u describes the vertical displacement of an elastic membrane spanned on a frame Γ under a density of force f and the assumption that the elastic energy $\left(\|\nabla u\|_{2,\Omega'}^2\right)$ remains bounded from above in Ω'.

Of course we picked up only a few examples. It is clear that many others can be obtained varying $a = a(u, v)$, f, V and considering intersections of convex sets K given by (4.17) or (4.22). We leave to the reader the pleasure to build his own collection of examples.

4.2. Quasilinear elliptic problems

4.2.1. Existence result. Let Ω be a bounded Lipschitz domain of $\mathbb{R}^n$. We consider for $i, j = 1, \ldots, n$ functions a_{ij}, β_i, a such that

$$a_{ij}, \beta_i, a : \Omega \times \mathbb{R} \to \mathbb{R} \text{ are Carathéodory functions.} \tag{4.32}$$

Recall that a function $A : \Omega \times \mathbb{R} \to \mathbb{R}$ is a Carathéodory function iff

$$\forall u \in \mathbb{R} \qquad x \mapsto A(x, u) \quad \text{is measurable,} \tag{4.33}$$

$$\text{a.e. } x \in \Omega \quad u \mapsto A(x, u) \quad \text{is continuous.} \tag{4.34}$$

Moreover we assume that for some $\lambda, \Lambda > 0$ it holds that

$$\lambda|\xi|^2 \le a_{ij}(x, u)\xi_i\xi_j \quad \text{a.e. } x \in \Omega, \quad \forall u \in \mathbb{R}, \quad \forall \xi \in \mathbb{R}^n, \tag{4.35}$$

$$\forall i, j = 1, \ldots, n : \ |a_{ij}(x, u)|, |\beta_i(x, u)|, |a(x, u)| \le \Lambda \text{ a.e. } x \in \Omega, \forall u \in \mathbb{R}. \tag{4.36}$$

Furthermore we suppose that for some function $a = a(x)$ it holds that

$$a(x, u) \ge a(x) \quad \text{a.e. } x \in \Omega, \quad \forall u \in \mathbb{R}. \tag{4.37}$$

Then, if V denotes the space $H^1(\Omega; \Gamma_0)$, if $f \in V'$ we would like to consider the class of problems

$$\begin{cases} u \in V, \\ \displaystyle\int_\Omega a_{ij}(x, u)\partial_{x_j}u\partial_{x_i}v + \beta_i(x, u)\partial_{x_i}v + a(x, u)uv \, dx = \langle f, v \rangle \quad \forall v \in V. \end{cases} \tag{4.38}$$

Such a problem is called "quasilinear". We can show

THEOREM 4.4. *Under the assumptions* (4.32)–(4.37) *and if in* (4.37) $a \not\equiv 0$ *when* $V = H^1(\Omega)$, *then for every* $f \in V'$ *there exists a solution to* (4.38).

PROOF. Let $w \in L^2(\Omega)$. By (4.32) we have

$$x \mapsto a_{ij}(x, w), \beta_i(x, w), a(x, w) \quad \text{are measurable.} \tag{4.39}$$

Moreover by (4.36) these functions are in $L^\infty(\Omega)$ and we have

$$\left| \int_\Omega \beta_i(x, w)\partial_{x_i}v \, dx \right| \le C\|\nabla v\|_{2,\Omega} \tag{4.40}$$

for some constant C. It follows that

$$v \mapsto \langle f, v \rangle - \int_\Omega \beta_i(x, w)\partial_{x_i}v \, dx \tag{4.41}$$

is an element of V'. By Theorem 1.8, (4.37) and the Lax–Milgram theorem it follows that there exists a solution $u = S(w)$ to

$$\begin{cases} \displaystyle\int_\Omega a_{ij}(x,w)\partial_{x_j}u\partial_{x_i}v + a(x,w)uv\,dx \\ \quad = \langle f,v\rangle - \displaystyle\int_\Omega \beta_i(x,w)\partial_{x_i}v\,dx \quad \forall v \in V, \\ u \in V. \end{cases} \tag{4.42}$$

It is clear that we will be done if we can show that $S : L^2(\Omega) \to L^2(\Omega)$ has a fixed point $u = S(u)$. For that, first we remark that there exists a constant C independent of w such that $|u|_{1,2} \leq C$. Indeed, choosing $v = u$ in the first equation of (4.42) we get – see (4.40)

$$\int_\Omega a_{ij}(x,w)\partial_{x_j}u\partial_{x_i}u + a(x,u)u^2\,dx \leq |f|_{V'}|u|_V + C||\nabla u||_{2,\Omega} \tag{4.43}$$

for some constant C. Using then (4.35), (4.37) it follows that

$$\lambda \int_\Omega |\nabla u|^2\,dx + \int_\Omega a(x)u^2\,dx \leq (|f|_{V'} + C)|u|_V$$

if we choose for instance $|u|_V = |u|_a$ – see (1.19). We derive then

$$\min(1,\lambda)|u|_a^2 \leq (|f|_{V'} + C)|u|_a \tag{4.44}$$

and the proof of the claim follows from Theorem 1.3. Thus if C denotes the constant such that

$$|u|_{1,2} \leq C, \tag{4.45}$$

consider S restricted to

$$B = B_C = \{v \in L^2(\Omega) \mid |v|_{2,\Omega} \leq C\}. \tag{4.46}$$

Clearly, S maps B into B. Moreover by (4.45) and Theorem 1.2, $S(B)$ is relatively compact in B. Thus, by the Schauder fixed point theorem (see [19]), we will be done provided we can show that S is continuous from B into B. This will follow from the claim

- S is continuous from $L^2(\Omega)$ into $L^2(\Omega)$.

Indeed, consider a sequence $w_n \in L^2(\Omega)$ such that

$$w_n \to w \quad \text{in} \quad L^2(\Omega). \tag{4.47}$$

By (4.45) it holds that – if $u_n = S(w_n)$

$$|u_n|_{1,2} \leq C$$

and for some subsequence n_k (see Theorem 1.2) we have for some $u_\infty \in H^1(\Omega)$

$$\begin{cases} u_{n_k} \rightharpoonup u_\infty & \text{in} & H^1(\Omega), \\ u_{n_k} \to u_\infty & \text{in} & L^2(\Omega), \\ u_{n_k} \to u_\infty & \text{a.e. in} & \Omega, \\ w_{n_k} \to w & \text{a.e. in} & \Omega. \end{cases} \tag{4.48}$$

Now u_{n_k} satisfies

$$\int_\Omega a_{ij}(x, w_{n_k})\partial_{x_j} u_{n_k} \partial_{x_i} v + a(x, w_{n_k})u_{n_k} v \, dx$$
$$= \langle f, v \rangle - \int_\Omega \beta_i(x, w_{n_k})\partial_{x_i} v \, dx \quad \forall v \in V. \tag{4.49}$$

Using the Lebesgue theorem it is easy to show that in $L^2(\Omega)$ it holds that

$$a_{ij}(x, w_{n_k})\partial_{x_i} v \quad \text{and} \quad a(x, w_{n_k})v \to a_{ij}(x, w)\partial_{x_i} v \quad \text{and} \quad a(x, w)v. \tag{4.50}$$

Then, passing to the limit in (4.49), using (4.48), we obtain for every $v \in V$

$$\int_\Omega a_{ij}(x, w)\partial_{x_j} u_\infty \partial_{x_i} v + a(x, w)u_\infty v \, dx$$
$$= \langle f, v \rangle - \int_\Omega \beta_i(x, w)\partial_{x_i} v \, dx. \tag{4.51}$$

Since $u_\infty \in V$ – see the first line of (4.48) – $u_\infty = u = S(w)$ and we have proved that

$$S(w_{n_k}) \to S(w) \quad \text{in} \quad L^2(\Omega). \tag{4.52}$$

If we did not have

$$S(w_n) \to S(w) \quad \text{in} \quad L^2(\Omega) \tag{4.53}$$

we would have for some $\varepsilon > 0$ and some subsequence

$$|S(w_{n_k}) - S(w)|_{2,\Omega} \geq \varepsilon. \tag{4.54}$$

But then, arguing as above, we would have (4.52) – eventually for some subsequence – and thus a contradiction to (4.54). Thus (4.53) holds and S is continuous. This completes the proof of the theorem.   □

4.2.2. Uniqueness results. In this section we would like to introduce a general method to prove uniqueness for the solution to problems of the type (4.38). The assumptions of Theorem 4.4 cannot guarantee uniqueness in general (see [3] for counterexamples). However controlling the modulus of continuity in u of the functions a_{ij}, β_i, a can lead to uniqueness. This is what we would like to analyze here.

We will suppose in this section that we are under the assumptions of Theorem 4.4 but with

$$a(x, u) = a(x) \tag{4.55}$$

i.e. this function is independent of u. We will say that a Carathéodory function $A(x, u)$ satisfies an H_w condition if

$$(H_\omega) \begin{cases} \text{there exists a function } w = \omega(t) \text{ on } [0, +\infty), \text{ nonnegative,} \\ \text{continuous and increasing such that} \\ |A(x, u) - A(x, v)| \leq w(|u - v|) \quad \forall u, v \in \mathbb{R}, \quad \text{a.e. } x \in \Omega. \end{cases}$$

Furthermore we will suppose that

$$\int_{0+} \frac{ds}{\omega^2(s)} = +\infty. \tag{4.56}$$

Such a condition is called an Osgood type condition. Note that it holds for instance when A is Hölder continuous in u uniformly in x with an exponent $\alpha \geq \frac{1}{2}$. We consider first the case where a is not degenerate, in other words the case where a is positive. Thus under the assumptions of Theorem 4.4 let u be the solution to

$$\begin{cases} u \in H_0^1(\Omega; \Gamma_0), \\ \displaystyle\int_\Omega a_{ij}(x, u)\partial_{x_j} u \partial_{x_i} v + \beta_i(x, u)\partial_{x_i} v + a(x)uv \, dx = \langle f, v \rangle \\ \qquad \forall v \in H_0^1(\Omega; \Gamma_0), \end{cases} \tag{4.57}$$

f is here some element of V' the dual space of $V = H_0^1(\Omega; \Gamma_0)$. Then, we can show

THEOREM 4.5. *Assume that we are under the assumptions of Theorem 4.4 in such a way that a solution to (4.57) exists for a satisfying $0 \leq a \leq \Lambda$ a.e. $x \in \Omega$. Furthermore let us assume that the functions*

$$a_{ij}(x, u), \quad \beta_i(x, u) \quad \text{satisfy} \quad (H_w) \quad \text{with (4.56).} \tag{4.58}$$

Finally we suppose that a is not degenerate in such a way that

$$a(x) > 0 \quad a.e. \quad x \in \Omega. \tag{4.59}$$

Then, under the above assumptions there exists a unique solution to (4.57).

PROOF. To simplify our notation let us introduce the functions

$$A_i(x, u, \xi) = a_{ij}(x, u)\xi_j + \beta_i(x, u) \tag{4.60}$$

(with of course the convention of summation in j in the right-hand side of (4.60), $x \in \Omega$, $u \in \mathbb{R}$, $\xi \in \mathbb{R}^n$.) With this in mind, the second equation of (4.57) reads

$$\int_\Omega A_i(x, u, \nabla u)\partial_{x_i} v + a(x)uv \, dx = \langle f, v \rangle \quad \forall v \in H_0^1(\Omega; \Gamma_0). \tag{4.61}$$

Then, let us first establish the following lemma:

LEMMA 4.6. *Let us assume that we are under the assumptions of Theorem 4.5. Let $u_1, u_2 \in H_0^1(\Omega; \Gamma_0)$ be two solutions to (4.61); then for any $\xi \in C^1(\overline{\Omega})$, $\xi \geq 0$ it holds that*

$$\int_{[u_2 - u_1 > 0]} \{A_i(x, u_2, \nabla u_2) - A_i(x, u_1, \nabla u_1)\}\partial_{x_i}\xi + a(x)(u_2 - u_1)\xi \, dx \leq 0. \tag{4.62}$$

PROOF. The method was introduced in [9]. First – replacing eventually w by a larger function – we can assume that

$$w(s) > 0 \quad \forall s > 0, \qquad \int^{+\infty} \frac{ds}{w^2(s)} < +\infty. \tag{4.63}$$

Thus for $\varepsilon > 0$ we can define

$$I_\varepsilon = \int_\varepsilon^{+\infty} \frac{ds}{\omega^2(s)} \tag{4.64}$$

and introduce a function F_ε by setting

$$F_\varepsilon(x) = \begin{cases} 0 & \text{if } x \leq \varepsilon, \\ \dfrac{1}{I_\varepsilon} \displaystyle\int_\varepsilon^x \frac{ds}{\omega^2(s)} & \text{if } x > \varepsilon. \end{cases} \tag{4.65}$$

Clearly, F_ε is a continuous differentiable function and it holds that

$$\xi F_\varepsilon(u_2 - u_1) \in H_0^1(\Omega; \Gamma_0). \tag{4.66}$$

Plugging this function in (4.61) written for u_1 and u_2 leads easily to

$$\int_\Omega A_i(x, u_2, \nabla u_2) \partial_{x_i} \{\xi F_\varepsilon(u_2 - u_1)\} - A_i(x, u_1, \nabla u_1) \partial_{x_i} \{\xi F_\varepsilon(u_2 - u_1)\} \\ + a(x)(u_2 - u_1) \xi F_\varepsilon(u_2 - u_1) \, dx = 0. \tag{4.67}$$

Performing the derivative ∂_{x_i} in the integral above leads to

$$\int_\Omega \{A_i(x, u_2, \nabla u_2) \partial_{x_i} \xi - A_i(x, u_1, \nabla u_1) \partial_{x_i} \xi\} F_\varepsilon(u_2 - u_1) \, dx \\ + \int_\Omega a(x)(u_2 - u_1) \xi F_\varepsilon(u_2 - u_1) \, dx \tag{4.68} \\ = \int_\Omega \{A_i(x, u_1, \nabla u_1) - A_i(x, u_2, \nabla u_2)\} \partial_{x_i} F_\varepsilon(u_2 - u_1) \xi \, dx.$$

Let us denote by I the right-hand side of (4.68). It holds that

$$I = \int_\Omega \{A_i(x, u_1, \nabla u_1) - A_i(x, u_1, \nabla u_2)\} \partial_{x_i} F_\varepsilon(u_2 - u_1) \xi \, dx \\ + \int_\Omega \{A_i(x, u_1, \nabla u_2) - A_i(x, u_2, \nabla u_2)\} \partial_{x_i} F_\varepsilon(u_2 - u_1) \xi \, dx. \tag{4.69}$$

By (4.65) we have

$$\partial_{x_i} F_\varepsilon(u_2 - u_1) = \chi_{[u_2 - u_1 > \varepsilon]} \cdot \frac{1}{I_\varepsilon} \cdot \frac{1}{w(u_2 - u_1)^2} \partial_{x_i}(u_2 - u_1). \tag{4.70}$$

$\chi_{[u_2 - u_1 > \varepsilon]}$ denotes the characteristic function of the set

$$[u_2 - u_1 > \varepsilon] = \{\, x \in \Omega \mid (u_2 - u_1)(x) > \varepsilon \,\}.$$

From (4.60) we derive then that

$$I = -\frac{1}{I_\varepsilon} \int_{[u_2-u_1>\varepsilon]} \frac{a_{ij}(x,u_1)\partial_{x_j}(u_2-u_1)\partial_{x_i}(u_2-u_1)\xi}{\omega^2(u_2-u_1)}\, dx$$

$$+ \frac{1}{I_\varepsilon} \int_{[u_2-u_1>\varepsilon]} \{a_{ij}(x,u_1) - a_{ij}(x,u_2)\}\frac{\partial_{x_j}u_2\partial_{x_i}(u_2-u_1)\xi}{\omega^2(u_2-u_1)}\, dx \qquad (4.71)$$

$$+ \frac{1}{I_\varepsilon} \int_{[u_2-u_1>\varepsilon]} \{\beta(x,u_1) - \beta(x,u_2)\}\frac{\partial_{x_i}(u_2-u_1)}{\omega^2(u_2-u_1)}\xi\, dx.$$

We can now use the assumption (H_ω) to get (see also (4.35))

$$I \leq \frac{1}{I_\varepsilon}\left\{ -\lambda \int_{[u_2-u_1>\varepsilon]} \frac{|\nabla(u_2-u_1)|^2}{\omega^2(u_2-u_1)}\xi\, dx \right.$$

$$+ C\int_{[u_2-u_1>\varepsilon]} |\nabla u_2|\frac{|\nabla(u_2-u_1)|}{\omega(u_2-u_1)}\xi\, dx \qquad (4.72)$$

$$\left. + C\int_{[u_2-u_1>\varepsilon]} \frac{|\nabla(u_2-u_1)|}{\omega(u_2-u_1)}\xi\, dx \right\}$$

for some constant C depending on n only.

By the Young inequality

$$ab \leq \frac{\lambda}{4C}a^2 + \frac{C}{\lambda}b^2$$

we get

$$I \leq \frac{1}{I_\varepsilon}\left\{ -\lambda \int_{[u_2-u_1>\varepsilon]} \frac{|\nabla(u_2-u_1)|^2}{\omega^2(u_2-u_1)}\xi\, dx \right.$$

$$+ C\left[\frac{\lambda}{4C}\int_{[u_2-u_1>\varepsilon]} \frac{|\nabla(u_2-u_1)|^2}{\omega^2(u_2-u_1)}\xi\, dx + \frac{C}{\lambda}\int_{[u_2-u_1>\varepsilon]} |\nabla u_2|^2\xi\, dx \right]$$

$$\left. + C\left[\frac{\lambda}{4C}\int_{[u_2-u_1>\varepsilon]} \frac{|\nabla(u_2-u_1)|^2}{\omega^2(u_2-u_1)}\xi\, dx + \frac{C}{\lambda}\int_{[u_2-u_1>\varepsilon]} \xi\, dx \right]\right\}.$$

From this it follows that

$$I \leq \frac{C^2}{\lambda I_\varepsilon}\left\{ \int_\Omega |\nabla u_2|^2\xi\, dx + \int_\Omega \xi\, dx \right\}.$$

Thus going back to (4.68) we have

$$\int_\Omega \{A_i(x,u_2,\nabla u_2) - A_i(x,u_1,\nabla u_1)\}\partial_{x_i}\xi F_\varepsilon(u_2-u_1)$$

$$+ a(x)(u_2-u_1)\xi F_\varepsilon(u_2-u_1)\, dx \qquad (4.73)$$

$$\leq \frac{C^2}{\lambda I_\varepsilon}\left\{ \int_\Omega (|\nabla u_2|^2+1)\xi\, dx \right\}.$$

By (4.56) it holds that

$$I_\varepsilon \to +\infty \qquad (4.74)$$

when $\varepsilon \to 0$. Moreover, since for $x \geq \varepsilon$ it holds that

$$F_\varepsilon(x) = \left\{ I_\varepsilon - \int_x^{+\infty} \frac{ds}{\omega^2(s)} \right\} / I_\varepsilon$$

and we have

$$F_\varepsilon(x) \to 0 \quad \text{for} \quad x \leq 0, \qquad F_\varepsilon(x) \to 1 \quad \text{for} \quad x > 0.$$

Thus, passing to the limit in (4.73) we obtain

$$\int_{[u_2 - u_1 > 0]} \{ A_i(x, u_2, \nabla u_2) - A_i(x, u_1, \nabla u_1) \} \partial_{x_i} \xi + a(u_2 - u_1) \xi \, dx \leq 0. \quad (4.75)$$

This completes the proof of the lemma. $\qquad\qquad\qquad\qquad\qquad\qquad\qquad \square$

END OF THE PROOF OF THEOREM 4.5. From (4.62) we derive that for any positive smooth function ξ it holds that

$$\int_{[u_2 - u_1 > 0]} \{ A_i(x, u_2, \nabla u_2) - A_i(x, u_1, \nabla u_1) \} \partial_{x_i} \xi \, dx \leq 0.$$

Replacing ξ by $M - \xi$ for $M \geq \xi$ we obtain

$$\int_{[u_2 - u_1 > 0]} \{ A_i(x, u_2, \nabla u_2) - A_i(x, u_1, \nabla u_1) \} \partial_{x_i} \xi \, dx = 0 \quad (4.76)$$

and from (4.62)

$$\int_{[u_2 - u_1 > 0]} a(u_2 - u_1) \xi \, dx \leq 0$$

for any smooth positive function ξ. Taking $\xi = 1$ implies that the set $[u_2 - u_1 > 0]$ has measure 0 and thus $u_2 \leq u_1$. Exchanging the rôles of u_1 and u_2 completes the proof of the theorem. $\qquad\qquad\qquad\qquad\qquad\qquad\qquad\qquad\qquad \square$

REMARK 4.1. In fact if u_1, u_2 are the solutions to (4.57) corresponding to $f_1, f_2 \in V'$ respectively we have established that

$$\langle f_1, v \rangle \geq \langle f_2, v \rangle \quad \forall v \in V, \quad v \geq 0 \quad (4.77)$$

implies that

$$u_2 \leq u_1 \quad (4.78)$$

i.e., the solution to (4.57) varies monotonically with respect to the data f.

It has been shown in [9] that uniqueness might fail in the case where (4.59) fails. However, in the case of Dirichlet boundary conditions uniqueness can be restored at the expense of monotonicity assumptions on the β_i. More precisely we consider u a solution to

$$\begin{cases} u \in H_0^1(\Omega), \\ \displaystyle\int_\Omega a_{ij}(x, u) \partial_{x_j} u \partial_{x_i} v + \beta_i(x, u) \partial_{x_i} v + a(x) u v \, dx = \langle f, v \rangle \; \forall v \in H_0^1(\Omega), \end{cases} \quad (4.79)$$

where $f \in H^{-1}(\Omega)$ the dual of $H_0^1(\Omega)$. Then we can show

THEOREM 4.7. *Assume that we are under the assumptions of Theorem 4.4 in such a way that a solution to (4.79) exists for*

$$0 \leq a(x) \leq \Lambda. \tag{4.80}$$

Moreover, let us assume that (4.58) holds. Furthermore suppose that there exist constants a_i not all equal to 0 such that

$$u \mapsto \sum_{i=1}^{n} a_i \beta_i(x, u) \tag{4.81}$$

is monotone a.e. $x \in \Omega$ (i.e., nonincreasing or nondecreasing). Then, when

$$|\partial_{x_j} a_{ij}(x, u)| \leq \Lambda' \quad \forall i, j = 1, \ldots, n \quad a.e. \ x \in \Omega, \quad \forall u \in \mathbb{R}, \tag{4.82}$$

(Λ' is some constant) the problem (4.79) admits a unique solution.

PROOF. We establish (4.76) exactly as in the previous theorem. Then, in (4.76) we choose $\xi = e^{\alpha(a,x)}$ where α will be chosen later on and (a, x) denotes the scalar product of $a = (a_1, \ldots, a_n)$ – see (4.81) – and x. Then (4.76) becomes:

$$\int_{[u_1-u_2>0]} \{a_{ij}(x, u_2)\partial_{x_j} u_2 - a_{ij}(x, u_1)\partial_{x_j} u_1\}\partial_{x_i} e^{\alpha(a,x)} \, dx$$

$$+ \int_{[u_2-u_1>0]} \sum_i a_i \{\beta_i(x, u_2) - \beta_i(x, u_1)\}\alpha e^{\alpha(a,x)} \, dx = 0.$$

We can choose the sign of α in such a way that the second integral above is nonpositive. Then, we get

$$\int_{[u_2-u_1>0]} \{a_{ij}(x, u_2)\partial_{x_j} u_2 - a_{ij}(x, u_1)\partial_{x_j} u_1\}\partial_{x_i} e^{\alpha(a,x)} \, dx \geq 0. \tag{4.83}$$

We define then

$$A_{ij}(x, u) = \int_0^u a_{ij}(x, s) \, ds. \tag{4.84}$$

Then, for $k = 1, 2$, $A_{ij}(x, u_k) \in H^1(\Omega)$ and we have

$$\partial_{x_j} A_{ij}(x, u_k) = a_{ij}(x, u_k)\partial_{x_j} u_k + \int_0^{u_k} \partial_{x_j} a_{ij}(x, s) \, ds. \tag{4.85}$$

Thus (4.83) becomes

$$\int_{[u_2-u_1>0]} \left\{\partial_{x_j}[A_{ij}(x, u_2) - A_{ij}(x, u_1)] - \int_{u_1}^{u_2} \partial_{x_j} a_{ij}(x, s) \, ds\right\}\partial_{x_i} e^{\alpha(a,x)} \, dx \geq 0.$$

Setting $w = (u_2 - u_1)^+$ we obtain

$$\int_\Omega \left[\partial_{x_j}[A_{ij}(x, u_1 + w) - A_{ij}(x, u_1)]\alpha a_i e^{\alpha(a,x)}\right.$$

$$\left. - \int_{u_1}^{u_1+w} \partial_{x_j} a_{ij}(x, s) \, ds \alpha a_i e^{\alpha(a,x)}\right] dx \geq 0.$$

Integrating by parts since $A_{ij}(x, u_1 + w) - A_{ij}(x, u_1) \in H_0^1(\Omega)$ we get by (4.84)

$$\int_\Omega \left[-\int_{u_1}^{u_1+w} a_{ij}(x,s)\, ds\, \alpha^2 a_i a_j e^{\alpha(a,x)} - \int_{u_1}^{u_1+w} \partial_{x_j} a_{ij}(x,s)\, ds\, \alpha a_i e^{\alpha(a,x)} \right] dx \geq 0.$$

Using the ellipticity condition we obtain that

$$\int_\Omega \int_{u_1}^{u_1+w} (-\lambda \alpha^2 |a|^2 + \Lambda' \alpha n |a|) e^{\alpha(a,x)}\, dx \geq 0.$$

If α is large enough we obtain a contradiction unless $w = 0$. This shows that $u_2 \leq u_1$ and completes the proof. (We can indeed exchange the rôles of u_1 and u_2.) $\quad\square$

4.3. Strongly nonlinear problems

We consider here mainly the question of uniqueness. Let u be a solution to the problem

$$\begin{cases} u \in H_0^1(\Omega; \Gamma_0), \\ \int_\Omega A_i(x, u, \nabla u) \partial_{x_i} v + a(x, u) v\, dx = \langle f, v \rangle \quad \forall v \in H_0^1(\Omega; \Gamma_0), \end{cases} \tag{4.86}$$

where f is an element of the dual of $H_0^1(\Omega; \Gamma_0)$, A_i is for $i = 1, \ldots, n$ a function from $\Omega \times \mathbb{R} \times \mathbb{R}^n$ of Carathéodory type – for instance the function given by (4.60) could be suitable.

We assume that there exists a function ω continuous, positive, nondecreasing on $(0, +\infty)$, $g \in L^2(\Omega)$, such that for $i = 1, \ldots, n$, $\xi \in \mathbb{R}^n$, $u \in \mathbb{R}$, a.e. $x \in \Omega$ it holds that

$$|A_i(x, u, \xi) - A_i(x, v, \xi)| \leq \omega(|u - v|)(|\xi| + g(x)) \tag{4.87}$$

with in addition ω satisfying

$$\int_{0+} \frac{ds}{\omega(s)} = +\infty. \tag{4.88}$$

((4.87), (4.88) hold for instance when $u \mapsto A_i(x, u, \xi)$ is Lipschitz continuous uniformly in x, ξ).

Furthermore we suppose that

$$\begin{cases} u \mapsto a(x, u) \text{ is nondecreasing}, \\ u \mapsto a(x, u) \text{ is increasing on } P \text{ in case } |\Gamma_0| = 0, \end{cases} \tag{4.89}$$

where P is a subset of Ω of positive measure and – with the summation convention

$$(A_i(x, u, \xi) - A_i(x, u, \xi'))(\xi_i - \xi_i') \geq \lambda |\xi - \xi'|^2 \tag{4.90}$$

for a.e. $x \in \Omega$, $\forall u \in \mathbb{R}$, $\forall \xi, \xi' \in \mathbb{R}^n$.

Then under these assumptions we can prove

THEOREM 4.8. *Under the above assumptions there is at most one solution to problem (4.86).*

PROOF. Let u_1, u_2 be two solutions to (4.86). Using as test function (see (4.65))

$$F_\varepsilon(u_2 - u_1) \in H_0^1(\Omega; \Gamma_0),$$

the second equation of (4.86) written for u_1 and u_2 leads to

$$\int_\Omega (A_i(x, u_2, \nabla u_2) - A_i(x, u_1, \nabla u_1)) \partial_{x_i} F_\varepsilon(u_2 - u_1)$$
$$+ (a(x, u_2) - a(x, u_1)) F_\varepsilon(u_2 - u_1) \, dx = 0.$$

Hence, we derive

$$\int_\Omega (A_i(x, u_2, \nabla u_2) - A_i(x, u_2, \nabla u_1)) \partial_{x_i}(u_2 - u_1) F_\varepsilon'(u_2 - u_1)$$
$$+ (a(x, u_2) - a(x, u_1)) F_\varepsilon(u_2 - u_1) \, dx$$
$$= \int_\Omega (A_i(x, u_1, \nabla u_1) - A_i(x, u_2, \nabla u_1)) \partial_{x_i}(u_2 - u_1) F_\varepsilon'(u_2 - u_1) \, dx.$$

Using (4.87), (4.90) we obtain

$$\lambda \int_\Omega F_\varepsilon'(u_2 - u_1) |\nabla(u_2 - u_1)|^2 \, dx + \int_\Omega (a(x, u_2) - a(x_1 u_1)) F_\varepsilon(u_2 - u_1) \, dx$$
$$\leq \int_\Omega \omega(|u_2 - u_1|) |\nabla(u_2 - u_1)| F_\varepsilon'(u_2 - u_1) (|\nabla u_1| + g) \, dx$$

and by the definition of F_ε we get

$$\frac{\lambda}{I_\varepsilon} \int_{[u_2 - u_1 > \varepsilon]} \frac{|\nabla(u_2 - u_1)|^2}{\omega^2(u_2 - u_1)} \, dx + \int_\Omega (a(x, u_2) - a(x, u_1)) F_\varepsilon(u_2 - u_1) \, dx$$
$$\leq \frac{1}{I_\varepsilon} \int_{[u_2 - u_1 > \varepsilon]} \frac{|\nabla(u_2 - u_1)|}{\omega(u_2 - u_1)} (|\nabla u_1| + g) \, dx.$$

It follows by applying in the last integral the Young inequality

$$ab \leq \frac{\lambda}{2} a^2 + \frac{1}{2\lambda} b^2$$

that

$$\frac{\lambda}{2 I_\varepsilon} \int_{[u_2 - u_1 > \varepsilon]} \frac{|\nabla(u_2 - u_1)|^2}{\omega^2(u_2 - u_1)} \, dx + \int_\Omega (a(x, u_2) - a(x, u_1)) F_\varepsilon(u_2 - u_1) \, dx$$
$$\leq \frac{1}{I_\varepsilon} \int_\Omega (|\nabla u_1| + g)^2 \, dx. \tag{4.91}$$

From (4.89) we derive after simplification by I_ε

$$\int_{[u_2 - u_1 > \varepsilon]} \frac{|\nabla(u_2 - u_1)|^2}{\omega(u_2 - u_1)^2} \leq \frac{2}{\lambda} \int_\Omega (|\nabla u_1| + g)^2 \, dx.$$

Setting

$$G_\varepsilon = \begin{cases} 0 & \text{if } x \le \varepsilon, \\ \displaystyle\int_\varepsilon^x \frac{ds}{\omega(s)} & \text{if } x > \varepsilon, \end{cases}$$

this reads also

$$\int_\Omega |\nabla G_\varepsilon(u_2 - u_1)|^2 \, dx \le C \tag{4.92}$$

where C is some constant independent of ε. In the case where $|\Gamma_0| = 0$, (4.91) implies also

$$\int_\Omega (a(x, u_2) - a(x, u_1)) F_\varepsilon(u_2 - u_1)) \, dx \le \frac{C}{I_\varepsilon} \to 0$$

when $\varepsilon \to 0$. Thus by (4.89), and since $F_\varepsilon(x) \to 1$ when $x > 0$, we have in this case

$$G_\varepsilon(u_2 - u_1) = 0 \quad \text{on } P. \tag{4.93}$$

It results from Theorem 1.5, 1.7, (4.92), (4.93) that in the case $|\Gamma_0| = 0$ it holds also as in the case $|\Gamma_0| \ne 0$

$$\int_\Omega G_\varepsilon(u_2 - u_1)^2 \le C$$

where C is independent of ε. Letting $\varepsilon \to 0$ and using (4.88) we get a contradiction unless $u_2 \le u_1$. The proof of $u_1 \ge u_2$ can be obtained by reversing the roles of u_1 and u_2. This completes the proof of the theorem.   □

REMARK 4.2. This theorem applies, under certain assumptions, to $A_i(x, u, \xi)$ given by (4.60).

Open problems

1. Show that the uniqueness result of Theorem 4.7 remains true under the same hypothesis for the solution u to

$$\begin{cases} u \in H_0^1(\Omega; \Gamma_0), \\ \displaystyle\int_\Omega a_{ij}(x, u)\partial_{x_j} u \partial_{x_i} v + \beta_i(x, u)\partial_{x_i} v + a(x)uv \, dx = \langle f, v \rangle \\ \forall v \in H_0^1(\Omega; \Gamma_0). \end{cases}$$

Here f denotes an element of the dual of $H_0^1(\Omega; \Gamma_0)$ (see also Remark 4.2).

2. Is it possible to relax the assumption (4.56) in Theorems 4.5, 4.7 ? – see also [**3**].

3. It would be interesting to consider the issues of uniqueness for variational inequalities associated to quasilinear operator and nonlocal constraints.

Chapter 5

Asymptotic Behaviour of some Nonlinear Elliptic Problems

In this chapter we will study the asymptotic behaviour of the nonlinear elliptic problems introduced in the preceding chapter. In comparison to the linear theory an additional difficulty arises here due to the possibility of multiple solutions to the problems at hand.

5.1. The case of variational inequalities

5.1.1. A simple example.

We consider the case where

$$\Omega_\ell = (-\ell, \ell) \times (-1, 1). \tag{5.1}$$

If $\omega = (-1, 1)$, let us denote by K the nonempty convex subset

$$K = \{\, v \in H_0^1(\omega) \mid v(x) \geq 0 \text{ a.e. in } \omega \,\}. \tag{5.2}$$

Then, according to Theorem 4.1, Proposition 4.2 – see also (4.26), (4.27) – for $f \in H^{-1}(\omega)$ there exists a unique u_∞ solution to

$$\begin{cases} \displaystyle\int_\omega \partial_{x_2} u_\infty \partial_{x_2}(v - u_\infty)\,dx_2 \geq \langle f, v - u_\infty \rangle \quad \forall v \in K, \\ u_\infty \in K. \end{cases} \tag{5.3}$$

Moreover, extending f to $H_0^1(\Omega_\ell)$ by the formula (3.21), there exists a unique solution u_ℓ to

$$\begin{cases} \displaystyle\int_{\Omega_\ell} \nabla u_\ell \nabla(v - u_\ell)\,dx \geq \langle f, v - u_\ell \rangle \quad \forall v \in K^\ell, \\ u_\ell \in K^\ell, \end{cases} \tag{5.4}$$

where K^ℓ is defined by

$$K^\ell = \{\, v \in H_0^1(\Omega_\ell) \mid v(x) \geq 0 \text{ a.e. } x = (x_1, x_2) \in \Omega_\ell \,\}. \tag{5.5}$$

(It results from Proposition 4.2 that K^ℓ is a closed convex set of $H_0^1(\Omega_\ell)$. Indeed take $K(x) = [0, +\infty) \times \mathbb{R}^2$. Moreover, as K, this convex set is not empty since it contains 0.)

Then we have

THEOREM 5.1. *For any $\ell_0 > 0$, $r > 0$ there exists a constant C independent of ℓ such that*

$$\left\|\nabla(u_\ell - u_\infty)\right\|_{2,\Omega_{\ell_0}} \leq \frac{C}{\ell^r}. \tag{5.6}$$

PROOF. Let $\ell_1 \leq \ell$ and let ϱ be the function defined by Figure 2.2. We claim that

$$u_\ell - (u_\ell - u_\infty)\varrho^2\left(\frac{x_1}{\ell_1}\right) \in K^\ell. \tag{5.7}$$

Indeed, it is clear that this function belongs to $H_0^1(\Omega_\ell)$. Moreover, since $\varrho \in [0,1]$ it holds that

$$u_\ell - (u_\ell - u_\infty)\varrho^2\left(\frac{x_1}{\ell_1}\right)$$
$$= \left(1 - \varrho^2\left(\frac{x_1}{\ell_1}\right)\right)u_\ell + \varrho^2\left(\frac{x_1}{\ell}\right)u_\infty \tag{5.8}$$
$$\geq 0.$$

This proves (5.7). Using this function as a test function in (5.4) we obtain

$$\int_{\Omega_\ell} \nabla u_\ell \nabla\left\{(u_\ell - u_\infty)\varrho^2\left(\frac{x_1}{\ell_1}\right)\right\}dx \leq \left\langle f, (u_\ell - u_\infty)\varrho^2\left(\frac{x_1}{\ell_1}\right)\right\rangle. \tag{5.9}$$

Recalling (3.21) we have

$$\left\langle f, (u_\ell - u_\infty)\varrho^2\left(\frac{x_1}{\ell_1}\right)\right\rangle = \int_{(-\ell,\ell)} \left\langle f, (u_\ell - u_\infty)(x_1, \cdot)\varrho^2\left(\frac{x_1}{\ell_1}\right)\right\rangle dx_1. \tag{5.10}$$

Moreover, for a.e. $x_1 \in (-\ell, \ell)$ it holds that

$$v = u_\infty + (u_\ell - u_\infty)(x_1, \cdot)\varrho^2\left(\frac{x_1}{\ell_1}\right) \in K. \tag{5.11}$$

Indeed, it results from Proposition 3.1 that a.e. $x_1 \in (-\ell, \ell)$

$$u_\infty + (u_\ell - u_\infty)(x_1, \cdot)\varrho^2\left(\frac{x_1}{\ell_1}\right) \in H_0^1(\omega).$$

Moreover, since $\varrho \in [0,1]$ it holds that

$$v = \left(1 - \varrho^2\left(\frac{x_1}{\ell_1}\right)\right)u_\infty + \varrho^2\left(\frac{x_1}{\ell_1}\right)u_\ell(x_1, \cdot)$$
$$\geq 0 \quad \text{a.e. } x_1 \in (-\ell, \ell)$$

(see (5.5)). Thus (5.11) holds. By (5.3) we then derive that

$$\int_\omega \partial_{x_2} u_\infty \partial_{x_2}(u_\ell - u_\infty)(x_1, x_2)\varrho^2\left(\frac{x_1}{\ell_1}\right)dx_2$$
$$\geq \left\langle f, (u_\ell - u_\infty)(x_1, \cdot)\varrho\left(\frac{x_1}{\ell_1}\right)\right\rangle \quad \text{a.e. } x_1 \in (-\ell, \ell). \tag{5.12}$$

Recalling (5.9), (5.10) we obtain

$$\int_{\Omega_\ell} \nabla u_\ell \nabla\left\{(u_\ell - u_\infty)\varrho^2\left(\frac{x_1}{\ell_1}\right)\right\}dx \leq \int_{\Omega_\ell} \partial_{x_2} u_\infty \partial_{x_2}(u_\ell - u_\infty)\varrho^2\left(\frac{x_1}{\ell_1}\right)dx. \tag{5.13}$$

Performing the derivations in the first integral and using the fact that u_∞ is independent of x_1 we deduce easily that

$$\int_{\Omega_{\ell_1}} |\nabla(u_\ell - u_\infty)|^2 \varrho^2\Big(\frac{x_1}{\ell_1}\Big) dx \leq -\frac{2}{\ell_1} \int_{\Omega_{\ell_1}} \partial_{x_1} u_\ell \cdot \partial_{x_1} \varrho\Big(\frac{x_1}{\ell_1}\Big)(u_\ell - u_\infty)\varrho\Big(\frac{x_1}{\ell}\Big) dx.$$

Arguing as in (2.16), (2.17) we get

$$\Big\| |\nabla(u_\ell - u_\infty)| \varrho\Big(\frac{x_1}{\ell_1}\Big) \Big\|^2_{2,\Omega_{\ell_1}} \leq \frac{4\sqrt{2}}{\ell_1} |\partial_{x_1} u_\ell|_{2,\Omega_{\ell_1}} \Big| \partial_{x_2}(u_\ell - u_\infty)\varrho\Big(\frac{x_1}{\ell_1}\Big) \Big|^2_{2,\Omega_{\ell_1}}.$$

From this we derive

$$\Big\| |\nabla(u_\ell - u_\infty)| \varrho\Big(\frac{x_1}{\ell_1}\Big) \Big\|^2_{2,\Omega_{\ell_1}} \leq \frac{4\sqrt{2}}{\ell_1} |\partial_{x_1} u_\ell|_{2,\Omega_{\ell_1}}. \tag{5.14}$$

Since u_∞ is independent of x_1 this implies (see (2.65)) that for any $\ell_1 \leq \ell$ it holds that

$$|\partial_{x_1} u_\ell|_{2,\Omega_{\ell_1}/2} \leq \frac{4\sqrt{2}}{\ell_1} |\partial_{x_1} u_\ell|_{2,\Omega_{\ell_1}}.$$

Applying this formula with $\ell_1 = \frac{\ell}{2^{k-2}}$, $\ell_1 = \frac{\ell}{2^{k-3}}, \ldots, \frac{\ell}{2}$ we derive easily for some constant $C = C_k$

$$|\partial_{x_1} u_\ell|_{2,\Omega_{\ell/2^{k-1}}} \leq \frac{C_k}{\ell^{k-1}} |\partial_{x_1} u_\ell|_{2,\Omega_\ell}. \tag{5.15}$$

Going back to (5.14) applied with $\ell_1 = \frac{\ell}{2^{k-1}}$, we obtain for some other constants C_k

$$\begin{aligned}
\big\| |\nabla(u_\ell - u_\infty)| \big\|_{2,\Omega_{\ell/2^k}} &\leq \Big\| |\nabla(u_\ell - u_\infty)| \varrho\Big(\frac{x_1}{\ell_1}\Big) \Big\|_{2,\Omega_{\ell_1}} \\
&\leq \frac{C_k}{\ell} |\partial_{x_1} u_\ell|_{2,\Omega_{\ell/2^{k-1}}} \tag{5.16} \\
&\leq \frac{C'_k}{\ell^k} |\partial_{x_1} u_\ell|_{2,\Omega_\ell}.
\end{aligned}$$

Thus, as before, in order to conclude we need an estimate of $|\partial_{x_1} u_\ell|_{2,\Omega_\ell}$. For that purpose in (5.4) we take $v = 0$. We obtain

$$\int_{\Omega_\ell} \nabla u_\ell \cdot \nabla(0 - u_\ell) \, dx \geq \langle f, 0 - u_\ell \rangle. \tag{5.17}$$

It follows that

$$\big\| \nabla u_\ell \big\|^2_{2,\Omega_\ell} \leq |f|_{*,\Omega_\ell} \big\| \nabla u_\ell \big\|_{2,\Omega_\ell} \tag{5.18}$$

where $|\cdot|_{*,\Omega_\ell}$ denotes the strong dual norm in $H^{-1}(\Omega_\ell)$. Recalling (3.20) we have

$$|f|_{*,\Omega_\ell} \leq \sqrt{2\ell} |f|_*$$

and from (5.18) we derive

$$\big\| \nabla u_\ell \big\|_{2,\Omega_\ell} \leq \sqrt{2\ell} |f|_*. \tag{5.19}$$

Going back to (5.16) it follows that

$$\big\| |\nabla(u_\ell - u_\infty)| \big\|_{2,\Omega_{\ell/2^k}} \leq \frac{C}{\ell^{k-\frac{1}{2}}} \tag{5.20}$$

where C is independent of ℓ. Choosing $k - \frac{1}{2} > r$, $\frac{\ell}{2^k} > \ell_0$, (5.6) follows. This completes the proof of the theorem. $\qquad\qquad\qquad\qquad\qquad\qquad\qquad\qquad\qquad\qquad\qquad\quad\square$

5.1.2. The general case. We suppose that

$$\Omega_\ell = (-\ell, \ell)^p \times \omega \tag{5.21}$$

where ω is a Lipschitz domain of $\mathbb{R}^{n-p}$. We adopt the notation of Chapter 3. In particular we will denote by $x = (X_1, X_2)$ the points of Ω_ℓ. We consider the bilinear form $a(u, v)$ defined by

$$a(u, v) = \int_{\Omega_\ell} a_{ij}(x)\partial_{x_j}u\partial_{x_i}v + a(X_2)uv \, dx \tag{5.22}$$

where the functions a_{ij}, a satisfy (3.3)–(3.6). We also assume that (3.8), (3.9) hold and set

$$V = H_0^1(\omega; \partial_0\omega). \tag{5.23}$$

In the case where $V = H^1(\omega)$ we suppose in addition that $a(X_2) \not\equiv 0$. If K is a closed convex subset of V, $K \neq \emptyset$ and $f \in V'$, it results then from Theorems 1.8, 4.1 that there exists a unique u_∞ solution to

$$\begin{cases} \displaystyle\int_\omega A_{22}\nabla_{X_2}u_\infty\nabla_{X_2}(v - u_\infty) + a(X_2)u_\infty(v - u_\infty)\, dX_2 \\ \qquad\qquad\qquad\qquad\qquad\qquad \geq \langle f, v - u_\infty\rangle \quad \forall v \in K, \\ u_\infty \in K. \end{cases} \tag{5.24}$$

(We adopted the notation (3.24)–(3.26).) We define then

$$\Gamma_0 = (-\ell, \ell)^p \times \partial_0\omega \tag{5.25}$$

and set

$$V_{\Omega_\ell} = H_0^1(\Omega_\ell; \Gamma_0). \tag{5.26}$$

It follows from our assumptions (in particular from the fact that $a \not\equiv 0$ when $|\partial_0\omega| = 0$) that $a(u, v)$ is a bilinear, continuous, coercive form on V_{Ω_ℓ}. We consider then for $\ell > 0$

$$K^\ell = \{\, v \in V_{\Omega_\ell} \mid v(X_1, \cdot) \in K \quad \text{a.e. } X_1 \in (-\ell, \ell)^p \,\}. \tag{5.27}$$

We have

PROPOSITION 5.2. K^ℓ is a nonempty closed convex subset of V_{Ω_ℓ}.

PROOF. If $v_0 \in K$, then the function

$$v_0(X_1, X_2) = v_0(X_2) \in K^\ell. \tag{5.28}$$

Indeed it is enough to show that $v_0 \in V_{\Omega_\ell}$, but this is clear. Thus, since K was assumed to be nonempty, so is K^ℓ.

Next, we consider a sequence v_n such that

$$v_n \in K^\ell, \qquad v_n \to v \quad \text{in } H^1(\Omega_\ell).$$

It is easy to see (cf. the proof of Proposition 3.1) that for a.e. $X_1 \in (-\ell, \ell)^p$ it holds that

$$v_n(X_1, \cdot) \to v(X_1, \cdot) \quad \text{in } H^1(\omega), \quad v_n(X_1, \cdot) \in K.$$

It follows – since K is closed in $H^1(\omega)$ – that

$$v(X_1, \cdot) \in K \quad \text{a.e. } X_1 \in (-\ell, \ell)^p$$

and this completes the proof of the proposition. $\qquad \square$

If $f \in V'$, we can define $f \in V'_{\Omega_\ell}$ by (3.21) and it is clear (see Theorem 4.1) that there exists a unique u_ℓ solution to

$$\begin{cases} \displaystyle\int_{\Omega_\ell} a_{ij}(x)\partial_{x_j} u_\ell \partial_{x_i}(v - u_\ell) + au_\ell(v - u_\ell)\, dx \\ \qquad\qquad\qquad\qquad \geq \langle f, v - u_\ell \rangle \quad \forall v \in K^\ell, \\ u_\ell \in K^\ell. \end{cases} \tag{5.29}$$

Moreover, we have

THEOREM 5.3. *Under the above assumptions, if u_ℓ is the solution to (5.29) and u_∞ the solution to (5.24), then $\forall \ell_0 > 0$, $\forall r > 0$ there exists a constant C independent of ℓ such that*

$$|u_\ell - u_\infty|_{V_{\Omega_{\ell_0}}} \leq \frac{C}{\ell^r}. \tag{5.30}$$

i.e. $u_\ell \to u_\infty$ in $H^1(\Omega_{\ell_0})$ with a speed higher than $\frac{1}{\ell^r}$ for any r.

PROOF. Let us denote by ϱ the function defined in (3.32), (3.33). Then for $\ell_1 \leq \ell$ we have

$$u_\ell - (u_\ell - u_\infty)\varrho^2\left(\frac{X_1}{\ell_1}\right) \in K^\ell. \tag{5.31}$$

(Recall that $\varrho \in [0, 1]$, and $u_\ell - (u_\ell - u_\infty)\varrho^2(\frac{X_1}{\ell_1}) = (1 - \varrho^2(\frac{X_1}{\ell_1}))u_\ell + \varrho^2(\frac{X_1}{\ell_1})u_\infty$.) Thus, from (5.29) we derive

$$\int_{\Omega_\ell} a_{ij}(x)\partial_{x_j} u_\ell \partial_{x_i}\left\{(u_\ell - u_\infty)\varrho^2\left(\frac{X_1}{\ell_1}\right)\right\} + au_\ell(u_\ell - u_\infty)\varrho^2\left(\frac{X_1}{\ell_1}\right) dx$$
$$\leq \left\langle f, (u_\ell - u_\infty)\varrho^2\left(\frac{X_1}{\ell_1}\right)\right\rangle. \tag{5.32}$$

Next, due to the definition of K^ℓ, for a.e. $X_1 \in (\ell, \ell)^p$ it holds that

$$u_\infty + (u_\ell - u_\infty)(X_1, \cdot)\varrho^2\left(\frac{X_1}{\ell_1}\right) \in K. \tag{5.33}$$

Thus from (5.24) we derive for a.e. $X_1 \in (-\ell, \ell)^p$

$$\int_\omega A_{22}(X_2)\nabla_{X_2} u_\infty \nabla_{X_2}(u_\ell - u_\infty)\varrho^2\left(\frac{X_1}{\ell_1}\right) + au_\infty(u_\ell - u_\infty)\varrho^2\left(\frac{X_1}{\ell_1}\right) dX_2$$
$$\geq \left\langle f, (u_\ell - u_\infty)(X_1, \cdot)\varrho^2\left(\frac{X_1}{\ell_1}\right)\right\rangle. \tag{5.34}$$

Integrating in X_1 and using (5.32) we derive

$$\int_{\Omega_\ell} A\nabla u_\ell \nabla \Big\{ (u_\ell - u_\infty)\varrho^2\Big(\frac{X_1}{\ell_1}\Big)\Big\} + au_\ell(u_\ell - u_\infty)\varrho^2\Big(\frac{X_1}{\ell_1}\Big)\, dx$$

$$\le \int_{\Omega_\ell} A_{22}\nabla_{X_2}u_\infty \nabla_{X_2}(u_\ell - u_\infty)\varrho^2\Big(\frac{X_1}{\ell_1}\Big) \tag{5.35}$$

$$+ au_\infty(u_\ell - u_\infty)\varrho^2\Big(\frac{X_1}{\ell_1}\Big)\, dx.$$

In other words in place of an equality in (3.34) we get an inequality. However we can argue exactly as in (3.35)–(3.37) in order to get (3.38), (3.39), that is to say

$$\int_{\Omega_{\ell_1}} \{|\nabla(u_\ell - u_\infty)|^2 + a(u_\ell - u_\infty)^2\}\varrho^2\Big(\frac{X_1}{\ell_1}\Big)\, dx$$

$$\le \frac{C}{\ell_1}\big\|\nabla(u_\ell - u_\infty)\big\|_{2,\Omega_{\ell_1}}\Big|(u_\ell - u_\infty)\varrho\Big(\frac{X_1}{\ell_1}\Big)\Big|_{2,\Omega_{\ell_1}}.$$

Applying Lemma 3.3 we get

$$\int_{\Omega_{\ell_1}} \{|\nabla(u_\ell - u_\infty)|^2 + a(u_\ell - u_\infty)^2\}\varrho^2\Big(\frac{X_1}{\ell_1}\Big)\, dx$$

$$\le \frac{C}{\ell_1}\big\|\nabla(u_\ell - u_\infty)\big\|_{2,\Omega_{\ell_1}}\Big(\int_{\Omega_{\ell_1}} \{|\nabla_{X_2}(u_\ell - u_\infty)|^2 \tag{5.36}$$

$$+ a(u_\ell - u_\infty)^2\}\varrho^2\Big(\frac{X_1}{\ell_1}\Big)dx\Big)^{1/2}.$$

Arguing as in (3.41), (3.42) we obtain for any $\ell_1 \le \ell$

$$\int_{\Omega_{\ell_1/2}} |\nabla(u_\ell - u_\infty)|^2 + a(u_\ell - u_\infty)^2\, dx \le \frac{C^2}{\ell_1^2}\int_{\Omega_{\ell_1}} |\nabla(u_\ell - u_\infty)|^2\, dx$$

and then we derive easily

$$\int_{\Omega_{\ell/2^k}} |\nabla(u_\ell - u_\infty)|^2 + a(u_\ell - u_\infty)^2\, dx \le \frac{C^2}{\ell^{2k}}\int_{\Omega_\ell} |\nabla(u_\ell - u_\infty)|^2\, dx \tag{5.37}$$

for some constant C. We have now to estimate this last integral. For that, considering $v_0 \in K$ and extending it to Ω_ℓ as in (5.28), we derive by (5.29)

$$\int_{\Omega_\ell} A\nabla u_\ell \nabla(v_0 - u_\ell) + au_\ell(v_0 - u_\ell)\, dx \ge \langle f, v_0 - u_\ell\rangle.$$

Thus we get

$$\int_{\Omega_\ell} A\nabla u_\ell \nabla u_\ell + au_\ell^2\, dx \le \int_{\Omega_\ell} A\nabla u_\ell \nabla v_0 + au_\ell v_0\, dx - \langle f, v_0 - u_\ell\rangle.$$

Using the ellipticity condition we obtain easily

$$\min(1, \lambda)\int_{\Omega_\ell} |\nabla u_\ell|^2 + au_\ell^2\, dx \le C|v_0|_{a,\Omega_\ell}|u_\ell|_{a,\Omega_\ell} + |\langle f, v_0 - u_\ell\rangle| \tag{5.38}$$

i.e.,

$$\min(1, \lambda)|u_\ell|^2_{a,\Omega_\ell} \leq C|v_0|_{a,\Omega_\ell}|u_\ell|_{a,\Omega_\ell} + |\langle f, v_0 \rangle| + |\langle f, u_\ell \rangle|.$$

It is clear that – if $|\cdot|_*$ is the strong dual norm on V' (see (5.19)) – we have for some constant C

$$|\langle f, v \rangle| \leq C\ell^{p/2}|f|_*|v|_{a,\Omega_\ell}$$

and we derive from above that

$$|u_\ell|^2_{a,\Omega_\ell} \leq C\{\ell^{p/2}|v_0|_{a,\omega}|u_\ell|_{a,\Omega_\ell} + \ell^p|f|_*|v_0|_{a,\omega} + \ell^{p/2}|f|_*|u_\ell|_{a,\Omega_\ell}\}.$$

Using the Young inequality it follows easily that

$$|u_\ell|^2_{a,\Omega_\ell} \leq C\ell^p\{|f|^2_* + |v_0|^2_{a,\omega}\}. \tag{5.39}$$

Going back to (5.37) we obtain then easily

$$|(u_\ell - u_\infty)|v_{\Omega_{\ell/2^k}} \leq \frac{C}{\ell^{k-\frac{p}{2}}} \tag{5.40}$$

for some constant C. One can then conclude as in the proof of Theorem 3.2. This completes the proof of the theorem. $\qquad \square$

EXAMPLES. In what follows $a(u, v)$ denotes the bilinear form defined by (5.22) and we set

$$a_\omega(u, v) = \int_\omega A_{22}(x_2)\nabla_{x_2}u\nabla_{x_2}v + a(x_2)uv\, dx_2. \tag{5.41}$$

To simplify the examples we will assume $f \in L^2(\omega)$ and consider Dirichlet boundary conditions. However, many other cases can of course be handled under the framework of Theorem 5.3.

1) The obstacle problem
Let $\varphi \in H_0^1(\omega)$. We set

$$\Gamma_0 = (-\ell, \ell)^p \times \partial\omega. \tag{5.42}$$

Then if u_ℓ denotes the solution to

$$\begin{cases} u_\ell \in K^\ell = \{v \in H_0^1(\Omega_\ell; \Gamma_0) \mid v \geq \varphi \text{ a.e. } x \in \Omega_\ell\}, \\ a(u_\ell, v - u_\ell) \geq \int_{\Omega_\ell} f(v - u_\ell)\, dx \quad \forall v \in K^\ell, \end{cases} \tag{5.43}$$

it holds that

$$u_\ell \to u_\infty \quad \text{in} \quad H^1(\Omega_{\ell_0})$$

where u_∞ is the solution to

$$\begin{cases} u_\infty \in K = \{v \in H_0^1(\omega) \mid v \geq \varphi \text{ a.e. } x_2 \in \omega\}, \\ a_\omega(u_\infty, v - u_\infty) \geq \int_\omega f(v - u_\infty)dx_2 \quad \forall v \in K. \end{cases} \tag{5.44}$$

Similarly, we could consider the case of the double obstacle problem.

2) The case of gradient constraints
Consider u_ℓ the solution to

$$\begin{cases} a(u_\ell, v - u_\ell) \geq \displaystyle\int_{\Omega_\ell} f(v - u_\ell)\,dx \quad \forall v \in K^\ell, \\ u_\ell \in K^\ell = \{\, v \in H_0^1(\Omega_\ell; \Gamma_0) \mid |\nabla_{X_2} v(x)| \leq 1 \text{ a.e. } x \in \Omega_\ell \,\}. \end{cases} \tag{5.45}$$

Then – see Remark 5.1 – u_ℓ converges in $H^1(\Omega_{\ell_0})$ toward u_∞ the solution to

$$\begin{cases} a_\omega(u_\infty, v - u_\infty) \geq \displaystyle\int_\omega f(v - u_\infty)\,dx \quad \forall v \in K, \\ u_\infty \in K = \{\, v \in H_0^1(\omega) \mid |\nabla_{X_2} v(X_2)| \leq 1 \text{ a.e. } X_2 \in \omega \,\}. \end{cases} \tag{5.46}$$

(Γ_0 is of course defined in (5.42).)

3) The case of nonlocal constraints for the limit problem
Consider u_ℓ the solution to

$$\begin{cases} u_\ell \in K^\ell = \left\{\, v \in H^1(\Omega_\ell; \Gamma_0) \;\middle|\; \displaystyle\int_\omega |\nabla_{X_2} v(X_1, X_2)|^2\,dX_2 \leq C \right. \\ \qquad\qquad\qquad\qquad\qquad\qquad\qquad \left. \text{a.e. } X_1 \in (-\ell, \ell)^p \right\}, \\ a(u_\ell, v - u_\ell) \geq \displaystyle\int_{\Omega_\ell} f(v - u_\ell)\,dx \quad \forall v \in K^\ell. \end{cases} \tag{5.47}$$

Then $u_\ell \to u_\infty$ in $H^{\frac{1}{2}}(\Omega_{\ell_0})$ where u_∞ is the solution to

$$\begin{cases} u_\infty \in K = \left\{\, v \in H_0^1(\omega) \;\middle|\; \displaystyle\int_\omega |\nabla_{X_2} v|^2\,dX_2 \leq C \right\}, \\ a_\omega(u_\infty, v - u_\infty) \geq \displaystyle\int_\omega f(v - u_\infty)\,dX_2 \quad \forall v \in K. \end{cases} \tag{5.48}$$

The convergence in the above examples follows directly from Theorem 5.3.

5.2. The case of quasilinear problems

As in the previous section we denote by Ω_ℓ the cylindrical domain defined by (5.1),

$$\Omega_\ell = (-\ell, \ell)^p \times \omega, \tag{5.49}$$

where ω is a bounded Lipschitz domain of $\mathbb{R}^{n-p}$, $1 \leq p < n$.

5.2.1. A first class of problems. We consider here some problems related to the linear case. More precisely, via a simple change of variable we will be able to derive the asymptotic behaviour of u_ℓ from the asymptotic behaviour described in Section 3.1 (see [13], [11]).

Consider for $i, j = 1, \ldots, p$ Carathéodory functions $a_{ij}(x, u)$ defined on $\mathbb{R}^p \times \omega \times \mathbb{R}$ i.e., such that

$$x \mapsto a_{ij}(x, u) \quad \text{is measurable } \forall u \in \mathbb{R}, \tag{5.50}$$

$$u \mapsto a_{ij}(x, u) \quad \text{is continuous a.e. } x \in \mathbb{R}^p \times \omega. \tag{5.51}$$

If $|\cdot|_\infty$ denotes the usual norm in $L^\infty(\mathbb{R}^p \times \omega)$ we will also assume that for some constant Λ it holds that

$$|a_{ij}(\cdot, u)|_\infty \leq \Lambda \quad \forall u \in \mathbb{R}, \quad \forall i, j = 1, \ldots, p. \tag{5.52}$$

We will in addition make the usual ellipticity assumption

$$\lambda|\xi|^2 \leq \sum_{i,j=1}^{p} a_{ij}(x, u)\xi_i\xi_j \quad \forall \xi \in \mathbb{R}^n, \ \forall u \in \mathbb{R}, \text{ a.e. } x \in \mathbb{R}^p \times \omega \tag{5.53}$$

where the above λ denotes a positive constant. We consider also a function b such that

$$b \text{ is continuous from } \mathbb{R} \text{ into } \mathbb{R}^+, \tag{5.54}$$

$$0 < a \leq b(u) \leq b \quad \forall u \in \mathbb{R}, \quad a, b \in \mathbb{R}. \tag{5.55}$$

As in the beginning of this chapter, points x in $\mathbb{R}^n$ are denoted by $x = (\mathrm{X}_1, \mathrm{X}_2)$ where $\mathrm{X}_1 \in \mathbb{R}^p$, $\mathrm{X}_2 \in \mathbb{R}^{n-p}$. Then for $\partial_0\omega$ a subset of $\partial\omega$ of positive measure we set

$$V = H_0^1(\omega; \partial_0\omega). \tag{5.56}$$

It follows from Theorem 4.4 that for $f \in V'$ there exists a solution u_∞ to

$$\begin{cases} u_\infty \in V, \\ \displaystyle\sum_{i=p+1}^{n} \int_\omega b(u_\infty)\partial_{x_i} u_\infty \partial_{x_i} v \, d\mathrm{X}_2 = \langle f, v \rangle \quad \forall v \in V. \end{cases} \tag{5.57}$$

Existence and uniqueness of a solution to (5.57) can be also obtained by noting that if

$$B(s) = \int_0^s b(t) \, dt \tag{5.58}$$

then $B(u_\infty)$ is the solution to the problem

$$\begin{cases} B(u_\infty) \in V, \\ \sum_{i=p+1}^{n} \int_\omega \partial_{x_i} B(u_\infty)\partial_{x_i} v \, d\mathrm{X}_2 = \langle f, v \rangle \quad \forall v \in V. \end{cases} \tag{5.59}$$

(Indeed, this follows from Lemma 1.16 since B is a function of class C^1 satisfying (1.83).) Having extended f by (3.21) and set as in (3.12)

$$\Gamma_0 = \partial(-\ell, \ell)^p \times \omega \cup (-\ell, \ell)^p \times \partial_0\omega \tag{5.60}$$

it follows from Theorem 4.4 that there exists a solution u_ℓ to

$$\begin{cases} u_\ell \in H_0^1(\Omega_\ell; \Gamma_0), \\ \displaystyle\sum_{i,j=1}^{p} \int_{\Omega_\ell} a_{ij}(x, u_\ell)\partial_{x_j} u_\ell \partial_{x_i} v \, dx \\ \quad + \displaystyle\sum_{i=p+1}^{n} \int_{\Omega_\ell} b(u_\ell)\partial_{x_i} u_\ell \partial_{x_i} v \, dx = \langle f, v \rangle \quad \forall v \in H_0^1(\Omega_\ell; \Gamma_0). \end{cases} \tag{5.61}$$

However, (see Chapter 4 or [**3**]) nothing insures the uniqueness of u_ℓ. Nevertheless we have

THEOREM 5.4. *Under the above assumptions – and if u_∞ denotes the unique solution to (5.57) – for any $\ell_0 > 0$ and any $r > 0$ there exists a constant C independent of ℓ such that*

$$|B(u_\ell) - B(u_\infty)|_{H^1(\Omega_{\ell_0})} \le \frac{C}{\ell^r}. \tag{5.62}$$

$(|\cdot|_{H^1(\Omega_{\ell_0})}$ *denotes a norm defining the usual topology of $H^1(\Omega_{\ell_0})$.)*

PROOF. We see – cf. again Lemma 1.16 – that $B(u_\ell)$ is a solution to

$$\begin{cases} B(u_\ell) \in H_0^1(\Omega_\ell; \Gamma_0), \\ \displaystyle\sum_{i,j=1}^p \int_{\Omega_\ell} \frac{a_{ij}(x, u_\ell)}{b(u_\ell)} \partial_{x_j} B(u_\ell) \partial_{x_i} v \, dx \\ \displaystyle + \sum_{i=p+1}^n \int_{\Omega_\ell} \partial_{x_i} B(u_\ell) \partial_{x_i} v \, dx = \langle f, v \rangle \quad \forall v \in H_0^1(\Omega_\ell; \Gamma_0). \end{cases} \tag{5.63}$$

Applying then Theorem 3.2 taking into account Remark 3.3 – the result follows. $\qquad\square$

As a corollary we have

COROLLARY 5.5. *Under the assumptions of Theorem 5.4, for any $\ell_0 > 0$ it holds that*

$$u_\ell \to u_\infty \quad in \ H^1(\Omega_{\ell_0}). \tag{5.64}$$

PROOF. This is an immediate consequence of the following lemma.

LEMMA 5.6. *Let Ω_0 be a bounded open subset of $\mathbb{R}^n$. Let $v_\ell, v_\infty \in H^1(\Omega_0)$. It holds that*

$$v_\ell \to v \quad in \ H^1(\Omega_0) \quad \Leftrightarrow \quad B(v_\ell) \to B(v) \quad in \ H^1(\Omega_0). \tag{5.65}$$

PROOF. It follows from Lemma 1.16 that all the above functions belong to $H^1(\Omega_0)$. Moreover, by (5.55), (5.58), we have

$$a|v_\ell - v_\infty| \le |B(v_\ell) - B(v_\infty)| = \left| \int_{v_\infty}^{v_\ell} b(s) \, ds \right| \le b|v_\ell - v_\infty| \tag{5.66}$$

which leads to

$$a|v_\ell - v_\infty|_{2,\Omega_0} \le |B(v_\ell) - B(v_\infty)|_{2,\Omega_0} \le b|v_\ell - v_\infty|_{2,\Omega_0}. \tag{5.67}$$

$(|\cdot|_{2,\Omega_0}$ is the usual norm in $L^2(\Omega_0)$.) Thus $B(v_\ell) \to B(v_\infty)$ in $L^2(\Omega_0)$ if and only if $v_\ell \to v_\infty$ in $L^2(\Omega_0)$. Next, for every i, it holds that

$$\begin{aligned} \partial_{x_i}(B(v_\ell) - B(v_\infty)) &= b(v_\ell)\partial_{x_i} v_\ell - b(v_\infty)\partial_{x_i} v_\infty \\ &= b(v_\ell)(\partial_{x_i} v_\ell - \partial_{x_i} v_\infty) + (b(v_\ell) - b(v_\infty))\partial_{x_i} u_\infty \end{aligned} \tag{5.68}$$

We claim that if $v_\ell - v_\infty \to 0$ in $L^2(\Omega_0)$, or $B(v_\ell) - B(v_\infty) \to 0$ in $L^2(\Omega_0)$ since it is equivalent, then

$$(b(v_\ell) - b(v_\infty))\partial_{x_i} v_\infty \to 0 \quad \text{in } L^2(\Omega_0). \tag{5.69}$$

Indeed, if not, then there exists a subsequence ℓ_k such that $\ell_k \to +\infty$ and for some ε

$$|(b(v_{\ell_k}) - b(v_\infty))\partial_{x_i} v_\infty|_{2,\Omega_0} \geq \varepsilon. \tag{5.70}$$

But from this sequence we can extract a subsequence – still labelled by k – such that

$$v_{\ell_k} \to v_\infty \quad \text{a.e. in } \Omega_0.$$

Then from (5.54), (5.55) we derive that

$$b(v_{\ell_k})\partial_{x_i} v_\infty \to b(v_\infty)\partial_{x_i} v_\infty \quad \text{a.e. in } \Omega_0,$$
$$|b(v_{\ell_k})\partial_{x_i} v_\infty| \leq b|\partial_{x_i} v_\infty| \quad \text{a.e. in } \Omega_0.$$

Since $b\partial_{x_i} v_\infty \in L^2(\Omega_0)$ it follows from the Lebesgue theorem that

$$|b(v_{\ell_k})\partial_{x_i} v_\infty - b(v_\infty)\partial_{x_i} v_\infty|_{2,\Omega_0} \to 0$$

which contradicts (5.70). Thus from (5.68), (5.69) we derive that

$$\partial_{x_i}(B(v_\ell) - B(v_\infty)) \to 0 \quad \text{in } L^2(\Omega_0)$$

if and only if

$$b(v_\ell)(\partial_{x_i} v_\ell - \partial_{x_i} v_\infty) \to 0 \quad \text{in } L^2(\Omega_0)$$

i.e., by (5.55) iff $\partial_{x_i} v_\ell - \partial_{x_i} v_\infty \to 0$ in $L^2(\Omega_0)$. This completes the proof of the lemma and of the corollary. $\qquad\square$

REMARK 5.1. From (5.62) we can also derive that

$$|u_\ell - u_\infty|_{2,\Omega_{\ell_0}} \leq \frac{C}{\ell^r}$$

where C is a constant independent of ℓ. We could also slightly generalize the technique replacing for instance the operator

$$-\sum_{i=p+1}^{n} \partial_{x_i}(b(u)\partial_{x_i} u)$$

by

$$-\sum_{i=p+1}^{n} \partial_{x_i}(a_i(\mathrm{x}_2)b(u)\partial_{x_i} u).$$

The details are left to the reader.

5.2.2. Another nonlinear case. In this section we will consider problems of the type introduced in Theorem 4.5. As seen in Section 5.2.1 the uniqueness of u_ℓ is not as essential as the uniqueness of u_∞. Thus we will only consider here situations where the limit solution is unique. As we are now used to, Ω_ℓ is defined by (5.49) and $x \in \mathbb{R}^n$ is written as (x_1, x_2). For $i, j = 1, \ldots, n$ we consider Carathéodory functions from $\mathbb{R}^p \times \omega \times \mathbb{R}$ onto $\mathbb{R}$, that is to say satisfying for every $i, j = 1, \ldots, n$

$$x \to a_{ij}(x, u) \quad \text{is measurable} \quad \forall u \in \mathbb{R}, \tag{5.71}$$

$$u \mapsto a_{ij}(x, u) \quad \text{is continuous} \qquad \text{a.e. } x \in \mathbb{R}^p \times \omega. \tag{5.72}$$

Moreover we will assume that for some positive constants λ, Λ it holds that

$$\lambda|\xi|^2 \le a_{ij}(x, u)\xi_i\xi_j \quad \text{a.e. } x \in \mathbb{R}^p \times \omega, \ \forall u \in \mathbb{R}, \ \forall \xi \in \mathbb{R}^n, \tag{5.73}$$

$$|a_{ij}(x, u)| \le \Lambda \quad \text{a.e. } x \in \mathbb{R}^p \times \omega, \ \forall u \in \mathbb{R}, \ \forall i, j = 1, \ldots, n. \tag{5.74}$$

As in Chapter 3 we will adopt the notation

$$A = \begin{pmatrix} A_{11} & A_{12} \\ A_{21} & A_{22} \end{pmatrix}$$

where the matrices A_{ij} are given by (3.25), (3.26). Similarly to (3.4) we will assume that

$$A_{12} = A_{12}(x, u) = A_{12}(x_2, u), \qquad A_{22} = A_{22}(x, u) = A_{22}(x_2, u) \tag{5.75}$$

i.e., A_{12} and A_{22} are independent of x_1. To be in the framework of Theorem 4.5 we introduce further Carathéodory functions $\beta_i(x, u)$, $i = p+1, \ldots, n$ satisfying

$$\beta_i(x, u) = \beta_i(x_2, u) \quad \forall i = p+1, \ldots, n, \tag{5.76}$$

$$|\beta_i(x_2, u)| \le \Lambda \quad \text{a.e. } x_2 \in \omega, \ \forall u \in \mathbb{R}, \ \forall i = p+1, \ldots, n \tag{5.77}$$

and a function $a \in L^\infty(\omega)$ satisfying

$$0 < \alpha \le a(x_2) \le \Lambda \tag{5.78}$$

for some positive constants α and Λ (note that without loss of generality the constant Λ in (5.74), (5.77) and (5.78) can be chosen to be the same).

Finally, since we would like the solution of the limit equation to be unique, we will suppose that there exists a constant γ and a nonnegative function $\omega : \mathbb{R}^+ \to \mathbb{R}^+$ satisfying

$$\omega(t) > 0 \quad \forall t > 0, \quad \omega \text{ nondecreasing}, \tag{5.79}$$

$$\int_{0+} \frac{ds}{\omega^2(s)} = +\infty, \tag{5.80}$$

such that $\forall i = 1, \ldots, n, \, j = p + 1, \ldots, n$ it holds that

$$|a_{ij}(\mathbf{x}_2, u) - a_{ij}(\mathbf{x}_2, v)| \leq \gamma \omega(|u - v|) \quad \text{a.e. } \mathbf{x}_2 \in \omega, \, \forall u, v \in \mathbb{R}, \tag{5.81}$$

$$|\beta_j(\mathbf{x}_2, u) - \beta_j(\mathbf{x}_2, v)| \leq \gamma \omega(|u - v|) \quad \text{a.e. } \mathbf{x}_2 \in \omega, \quad \forall u, v \in \mathbb{R}. \tag{5.82}$$

Then, according to Theorem 4.5, applied possibly with ω replaced by $\gamma \omega$, for $f \in H^{-1}(\omega)$ there exists a unique u_∞ solution to

$$\begin{cases} u_\infty \in H_0^1(\omega), \\ \displaystyle \int_\omega A_{22}(\mathbf{x}_2, u_\infty) \nabla_{\mathbf{x}_2} u_\infty \nabla_{\mathbf{x}_2} v + \beta_j(\mathbf{x}_2, u_\infty) \partial_{x_j} v + a(\mathbf{x}_2) u_\infty v \, d\mathbf{x}_2 \\ \hspace{4cm} = \langle f, v \rangle \quad \forall v \in H_0^1(\omega). \end{cases} \tag{5.83}$$

(In the last equation the summation in j is taken from $p + 1$ to n, recall that the vector $\nabla_{\mathbf{x}_2}$ is the vector $(\partial_{x_{p+1}}, \ldots, \partial_{x_n})$.)

Moreover, see Theorem 4.4, for every $\ell > 0$ there exists u_ℓ solution to

$$\begin{cases} u_\ell \in H_0^1(\Omega_\ell), \\ \displaystyle \int_{\Omega_\ell} A(x, u_\ell) \nabla u_\ell \nabla v + \beta_j(\mathbf{x}_2, u_\ell) \partial_{x_j} v + a(\mathbf{x}_2) u_\ell v \, dx \\ \hspace{4cm} = \langle f, v \rangle \quad \forall v \in H_0^1(\Omega_\ell). \end{cases} \tag{5.84}$$

As usual in the above formula the bracket $\langle f, v \rangle$ is defined by (3.21). The summation in j is from $p + 1$ to n exactly as in (5.83). Of course we do not know if in general the solution to (5.84) is unique. However, as we will see, the asymptotic behaviour of such a u_ℓ is perfectly determined. Let us first prove

THEOREM 5.7. *Under the above assumptions let u_ℓ be a solution to* (5.84) *and u_∞ be the solution to* (5.83). *Assume further that for some constant Λ it holds that*

$$|\partial_{x_k} a_{ij}(x, u)| \leq \Lambda \quad \text{a.e. } x \in \mathbb{R}^p \times \omega, \, \forall u \in \mathbb{R} \, \forall i, j, k = 1, \ldots, n. \tag{5.85}$$

Then for any $\ell_0 > 0$, $r > 0$ there exists a constant C independent of ℓ such that

$$\int_{\Omega_{\ell_0}} |u_\ell - u_\infty| \, dx \leq \frac{C}{\ell^r}. \tag{5.86}$$

Before giving the proof of the theorem, let us first prove the following estimate for u_ℓ solution to (5.84).

LEMMA 5.8. *Let u_ℓ be a solution to* (5.84); *then there exists $C = C(\omega, \lambda, \Lambda, n, p, f)$ independent of ℓ such that*

$$\|\nabla u_\ell\|_{2, \Omega_\ell} \leq C \ell^{\frac{p}{2}}. \tag{5.87}$$

PROOF. If $|\cdot|_*$ denotes the strong dual norm in $H^{-1}(\omega)$ when $H_0^1(\omega)$ is equipped of the norm $||\nabla v||_{2,\omega}$, it follows as in (3.20) that

$$|\langle f, v \rangle| \leq (2\ell)^{\frac{p}{2}} |f|_* ||\nabla v||_{2,\Omega_\ell}. \tag{5.88}$$

Taking then $v = u_\ell$ in (5.84) we obtain (see (5.77))

$$\int_{\Omega_\ell} A \nabla u_\ell \nabla u_\ell + a(x_2) u_\ell^2 \, dx$$

$$\leq |\langle f, u_\ell \rangle| + \sum_{j=p+1}^{n} \int_{\Omega_\ell} |\beta_j(x_2, u_\ell)| |\nabla u_\ell| \, dx$$

$$\leq (2\ell)^{\frac{p}{2}} |f|_* ||\nabla u_\ell||_{2,\Omega_\ell} + (n-p)\Lambda \int_{\Omega_\ell} |\nabla u_\ell| \, dx$$

$$\Rightarrow \quad \lambda ||\nabla u_\ell||_{2,\Omega_\ell}^2 \leq (2\ell)^{\frac{p}{2}} |f|_* ||\nabla u_\ell||_{2,\Omega_\ell} + (n-p)\Lambda(2\ell)^{\frac{p}{2}} |\omega|^{\frac{1}{2}} ||\nabla u_\ell||_{2,\Omega_\ell}.$$

The inequality (5.87) follows then easily. $\qquad\square$

We turn now to the proof of Theorem 5.7.

PROOF OF THEOREM 5.7. Let us consider $\varrho = \varrho(x_1)$ a smooth function such that

$$0 \leq \varrho \leq 1, \qquad \varrho = 1 \text{ on } \left(-\frac{1}{2}, \frac{1}{2}\right)^p, \qquad \varrho = 0 \text{ outside of } (-1,1)^p, \tag{5.89}$$

$$|\nabla_{x_1} \varrho| \leq \theta \tag{5.90}$$

where θ is some constant. Moreover, let us consider F_ε the function defined by (4.63)–(4.65). It is clear that for any $\ell_1 \leq \ell$ the function

$$F_\varepsilon(u_\ell - u_\infty)\varrho^2\left(\frac{x_1}{\ell_1}\right) \in H_0^1(\Omega_\ell) \tag{5.91}$$

and for a.e. x_1

$$F_\varepsilon(u_\ell - u_\infty)(x_1, \cdot)\varrho^2\left(\frac{x_1}{\ell_1}\right) \in H_0^1(\omega). \tag{5.92}$$

Using these two functions in (5.84), (5.83) respectively – and denoting (5.91) by $F_\varepsilon \varrho^2$ – we obtain

$$\int_{\Omega_\ell} A \nabla u_\ell \nabla(F_\varepsilon \varrho^2) + \beta_j(x_2, u_\ell)\partial_j(F_\varepsilon \varrho^2) + a u_\ell F_\varepsilon \varrho^2 \, dx$$

$$= \int_{\Omega_\ell} A_{22} \nabla_{x_2} u_\infty \nabla_{x_2}(F_\varepsilon \varrho^2) + \beta_j(x_2, u_\infty)\partial_{x_j}(F_\varepsilon \varrho^2) + a u_\infty F_\varepsilon \varrho^2 \, dx. \tag{5.93}$$

(Note that the summation in j in the above formula is between $p+1$ and n.)

Expanding the first integral above we obtain (recall (5.75))

$$
\int_{\Omega_\ell} A \nabla u_\ell \nabla (F_\varepsilon \varrho^2) \, dx - \int_{\Omega_\ell} A_{22}(\mathrm{x}_2, u_\infty) \nabla_{\mathrm{x}_2} u_\infty \nabla_{\mathrm{x}_2} (F_\varepsilon \varrho^2)
$$

$$
= \int_{\Omega_\ell} A_{11}(x, u_\ell) \nabla_{\mathrm{x}_1} u_\ell \nabla_{\mathrm{x}_1} (F_\varepsilon \varrho^2) \, dx
$$

$$
+ \int_{\Omega_\ell} (A_{12}(\mathrm{x}_2, u_\ell) \nabla_{\mathrm{x}_2} u_\ell - A_{12}(\mathrm{x}_2, u_\infty) \nabla_{\mathrm{x}_2} u_\infty) \nabla_{\mathrm{x}_1} (F_\varepsilon \varrho^2) \, dx
$$

$$
+ \int_{\Omega_\ell} A_{21}(x, u_\ell) \nabla_{\mathrm{x}_1} u_\ell \nabla_{\mathrm{x}_2} (F_\varepsilon \varrho^2) \, dx \tag{5.94}
$$

$$
+ \int_{\Omega_\ell} (A_{22}(\mathrm{x}_2, u_\ell) \nabla_{\mathrm{x}_2} u_\ell - A_{22}(\mathrm{x}_2, u_\infty) \nabla_{\mathrm{x}_2} u_\infty) \nabla_{\mathrm{x}_2} (F_\varepsilon \varrho^2)
$$

$$
+ \int_{\Omega_\ell} A_{12}(\mathrm{x}_2, u_\infty) \nabla_{\mathrm{x}_2} u_\infty \nabla_{\mathrm{x}_1} (F_\varepsilon \varrho^2) \, dx.
$$

(This last integral vanishes since $A_{12}(\mathrm{x}_2, u_\infty) \nabla_{\mathrm{x}_2} u_\infty$ is independent of x_1.) Combining (5.93), (5.94) we obtain

$$
E \stackrel{\text{Def}}{=} \int_{\Omega_\ell} A_{11}(x, u_\ell) \nabla_{\mathrm{x}_1} u_\ell \nabla_{\mathrm{x}_1} \varrho^2 F_\varepsilon \, dx
$$

$$
+ \int_{\Omega} (A_{12}(\mathrm{x}_2, u_\ell) \nabla_{\mathrm{x}_2} u_\ell - A_{12}(\mathrm{x}_2, u_\infty) \nabla_{\mathrm{x}_2} u_\infty) \nabla_{\mathrm{x}_1} \varrho^2 F_\varepsilon \, dx
$$

$$
+ \int_{\Omega_\ell} a(u_\ell - u_\infty) F_\varepsilon \varrho^2
$$

$$
= - \int_{\Omega_\ell} A_{11}(x, u_\ell) \nabla_{\mathrm{x}_1} u_\ell \nabla_{\mathrm{x}_1} F_\varepsilon \varrho^2 \, dx
$$

$$
- \int_{\Omega_\ell} (A_{12}(\mathrm{x}_2, u_\ell) \nabla_{\mathrm{x}_2} u_\ell - A_{12}(\mathrm{x}_2, u_\infty) \nabla_{\mathrm{x}_2} u_\infty) \nabla_{\mathrm{x}_2} F_\varepsilon \varrho^2 \, dx \tag{5.95}
$$

$$
- \int_{\Omega_\ell} A_{21}(x, u_\ell) \nabla_{\mathrm{x}_1} u_\ell \nabla_{\mathrm{x}_1} F_\varepsilon \varrho^2 \, dx
$$

$$
- \int_{\Omega_\ell} (A_{22}(\mathrm{x}_2, u_\ell) \nabla_{\mathrm{x}_2} u_\ell - A_{22}(\mathrm{x}_2, u_\infty) \nabla_{\mathrm{x}_2} u_\infty) \nabla_{\mathrm{x}_2} F_\varepsilon \varrho^2 \, dx
$$

$$
- \int_{\Omega_\ell} (\beta_j(\mathrm{x}_2, u_\ell) - \beta_j(\mathrm{x}_2, u_\infty)) \partial_{x_j} F_\varepsilon \varrho^2 \, dx.
$$

(Recall that the summation in j is for $j > p$.) Using the fact that u_∞ is independent of x_1, and the equality

$$
A_{i2}(\mathrm{x}_2, u_\ell) \nabla_{\mathrm{x}_2} u_\ell - A_{i2}(\mathrm{x}_2, u_\infty) \nabla_{\mathrm{x}_2} u_\infty
$$

$$
= A_{i2}(\mathrm{x}_2, u_\ell) \nabla_{\mathrm{x}_2} (u_\ell - u_\infty) + (A_{i2}(\mathrm{x}_2, u_\ell) - A_{i2}(\mathrm{x}_2, u_\infty)) \nabla_{\mathrm{x}_2} u_\infty
$$

we obtain easily

$$E = -\int_{\Omega_\ell} A(x, u_\ell) \nabla(u_\ell - u_\infty) \nabla F_\varepsilon \varrho^2 \, dx$$

$$-\sum_{i=1}^{2} \int_{\Omega_\ell} (A_{i2}(\mathbf{x}_2, u_\ell) - A_{i2}(\mathbf{x}_2, u_\infty)) \nabla_{\mathbf{x}_2} u_\infty \nabla_{\mathbf{x}_i} F_\varepsilon \varrho^2 \, dx$$

$$-\int_{\Omega_\ell} (\beta_j(\mathbf{x}_2, u_\ell) - \beta_j(\mathbf{x}_2, u_\infty)) \partial_{x_j} F_\varepsilon \varrho^2 \, dx.$$

We have

$$\nabla F_\varepsilon = \nabla F_\varepsilon(u_\ell - u_\infty) = \frac{1}{I_\varepsilon} \chi_{[u_\ell - u_\infty > \varepsilon]} \frac{1}{\omega^2(u_\ell - u_\infty)} \nabla(u_\ell - u_\infty)$$

(see (4.70)). Thus we derive from (5.73), (5.81), (5.82)

$$E \le \frac{1}{I_\varepsilon} \left\{ -\lambda \int_{[u_\ell - u_\infty > \varepsilon]} \frac{|\nabla(u_\ell - u_\infty)|^2}{\omega^2} \varrho^2 \right.$$

$$+ C\gamma \int_{[u_\ell - u_\infty > \varepsilon]} \omega |\nabla_{\mathbf{x}_2} u_\infty| \frac{|\nabla(u_\ell - u_\infty)|}{\omega^2} \varrho^2 \, dx$$

$$\left. + C\gamma \int_{[u_\ell - u_\infty > \varepsilon]} \omega \frac{|\nabla(u_\ell - u_\infty)|}{\omega^2} \varrho^2 \, dx \right\}.$$

In the above inequality C is some constant depending on n only, ω stands for $\omega(u_\ell - u_\infty)$, $[u_\ell - u_\infty > \varepsilon] = \{ x \in \Omega_\ell \mid (u_\ell - u_\infty)(x) > \varepsilon \}$. Denoting by Ω_ε this last set we obtain

$$E \le \frac{1}{I_\varepsilon} \left\{ -\lambda \int_{\Omega_\varepsilon} \frac{|\nabla(u_\ell - u_\infty)|^2}{\omega^2} \varrho^2 \, dx \right.$$

$$\left. + C\gamma \int_{\Omega_\varepsilon} (|\nabla_{\mathbf{x}_2} u_\infty| + 1) \frac{|\nabla(u_\ell - u_\infty)|}{\omega} \varrho^2 \, dx \right\}.$$

Using in this last integral the Young inequality

$$ab \le \frac{\lambda}{2C\gamma} a^2 + \frac{C\gamma}{2\lambda} b^2$$

we obtain

$$E \le \frac{1}{I_\varepsilon} \left\{ -\lambda \int_{\Omega_\varepsilon} \frac{|\nabla(u_\ell - u_\infty)|^2}{\omega^2} \varrho^2 \, dx + \frac{\lambda}{2} \int_{\Omega_\varepsilon} \frac{|\nabla(u_\ell - u_\infty)|^2}{\omega^2} \varrho^2 \, dx \right.$$

$$\left. + \frac{C^2\gamma^2}{2\lambda} \int_{\Omega_\varepsilon} (|\nabla_{\mathbf{x}_2} u_\infty| + 1)^2 \varrho^2 \, dx \right\} \qquad (5.96)$$

$$\le \frac{1}{I_\varepsilon} \frac{C^2\gamma^2}{2\lambda} \int_{\Omega_\varepsilon} (|\nabla_{\mathbf{x}_2} u_\infty| + 1)^2 \varrho^2 \, dx.$$

Recalling the expression of E – see (5.95) – and the fact that (see (4.74) and below)

$$I_\varepsilon \to +\infty, \qquad F_\varepsilon \to \chi_{(0, +\infty)}$$

we obtain, passing to the limit in (5.96),

$$\int_{[u_\ell - u_\infty > 0]} a(u_\ell - u_\infty)\varrho^2 \, dx$$

$$\leq - \int_{[u_\ell - u_\infty > 0]} A_{11}(x, u_\ell)\nabla_{X_1} u_\ell \nabla_{X_1}\varrho^2 \, dx \tag{5.97}$$

$$- \int_{[u_\ell - u_\infty > 0]} (A_{12}(X_2, u_\ell)\nabla_{X_2} u_\ell - A_{12}(X_2, u_\infty))\nabla_{X_2} u_\infty \nabla_{X_1}\varrho^2 \, dx.$$

We then set

$$\tilde{a}_{ij}(x, z) = \int_0^z a_{ij}(x, s) \, ds. \tag{5.98}$$

For $u \in H^1(\Omega_\ell)$ it holds that

$$\tilde{a}_{ij}(x, u) \in H^1(\Omega_\ell)$$

and for every $k = 1, \ldots, n$

$$\partial_{x_k}\tilde{a}_{ij}(x, u) = a_{ij}(x, u)\partial_{x_k} u + \int_0^u \partial_{x_k} a_{ij}(x, s) \, ds. \tag{5.99}$$

Using this in (5.97) we get

$$\int_{[u_\ell - u_\infty > 0]} a(u_\ell - u_\infty)\varrho^2 \, dx$$

$$\leq - \int_{[u_\ell - u_\infty > 0]} \sum_{i,j=1}^p \left\{ \partial_{x_j}\tilde{a}_{ij}(x, u_\ell) - \int_0^{u_\ell} \partial_{x_j} a_{ij}(x, s) \, ds \right\} \partial_{x_i}\varrho^2 \, dx$$

$$- \int_{[u_\ell - u_\infty > 0]} \sum_{\substack{i=1,\ldots,p \\ j=p+1,\ldots,n}} \left\{ \partial_{x_j}(\tilde{a}_{ij}(X_2, u_\ell) - \tilde{a}_{ij}(X_2, u_\infty)) \right. \tag{5.100}$$

$$\left. - \int_{u_\infty}^{u_\ell} \partial_{x_j} a_{ij}(X_2, s) \, ds \right\} \partial_{x_i}\varrho^2 \, dx.$$

For $j = 1, \ldots, p$ we derive from (5.99) that

$$\partial_{x_j}\tilde{a}_{ij}(x, u_\infty) = \int_0^{u_\infty} \partial_{x_j} a_{ij}(x, s) \, ds$$

and (5.100) becomes

$$\int_{[u_\ell - u_\infty > 0]} a(u_\ell - u_\infty)\varrho^2 \, dx$$

$$\leq - \sum_{\substack{i=1,\ldots,p \\ j=1,\ldots,n}} \int_{[u_\ell - u_\infty > 0]} \left\{ \partial_{x_j}(\tilde{a}_{ij}(x, u_\ell) - \tilde{a}_{ij}(x, u_\infty)) \right. \tag{5.101}$$

$$\left. - \int_{u_\infty}^{u_\ell} \partial_{x_j} a_{ij}(x, s) \, ds \right\} \partial_{x_i}\varrho^2 \, dx.$$

We then set $w = (u_\ell - u_\infty)^+$. The above inequality implies that

$$\int_{\Omega_\ell} aw\varrho^2 \, dx \leq - \sum_{\substack{i=1,\ldots,p \\ j=1,\ldots,n}} \int_{\Omega_\ell} \Big\{ \partial_{x_j} \big(\tilde{a}_{ij}(x, u_\infty + w) - \tilde{a}_{ij}(x, u_\infty) \big)$$
$$- \int_{u_\infty}^{u_\infty + w} \partial_{x_j} a_{ij}(x, s) \, ds \Big\} \partial_{x_i} \varrho^2 \, dx.$$

Integrating by parts we obtain

$$\int_{\Omega_\ell} aw\varrho^2 \, dx \leq \sum_{\substack{i=1,\ldots,p \\ j=1,\ldots,n}} \int_{\Omega_\ell} \{ \tilde{a}_{ij}(x, u_\infty + w) - \tilde{a}_{ij}(x, u_\infty) \} \partial^2_{x_i x_j} \varrho^2$$
$$- \int_{u_\infty}^{u_\infty + w} \partial_{x_j} a_{ij}(x, s) \, ds \partial_{x_i} \varrho^2 \, dx. \tag{5.102}$$

Using the definition of $\tilde{a}_{ij}$ we have

$$\tilde{a}_{ij}(x, u_\infty + w) - \tilde{a}_{ij}(x, u_\infty) = \int_{u_\infty}^{u_\infty + w} a_{ij}(x, s) \, ds.$$

Thus we obtain for a constant $C = C(n, \Lambda)$ (see (5.85))

$$\int_{\Omega_\ell} aw\varrho^2 \, dx \leq C \sum_{\substack{i=1,\ldots,p \\ j=1,\ldots,n}} \int_{\Omega_\ell} w\{ |\partial^2_{x_i x_j} \varrho^2| + \partial_{x_i} \varrho^2| \} \, dx. \tag{5.103}$$

We have for $i = 1, \ldots, p$, $j = 1, \ldots, n$

$$\left| \partial_{x_i} \varrho^2 \left(\frac{\mathrm{x}_1}{\ell_1} \right) \right| = \left| 2\partial_{x_i} \varrho \left(\frac{\mathrm{x}_1}{\ell_1} \right) \varrho \left(\frac{\mathrm{x}_1}{\ell_1} \right) \frac{1}{\ell_1} \right| \leq \frac{C}{\ell_1}$$

$$\left| \partial^2_{x_i x_j} \varrho^2 \left(\frac{\mathrm{x}_1}{\ell_1} \right) \right| \leq \frac{C}{\ell_1}$$

where C is a constant independent of $\ell_1 \geq 1$. Since in the integrals of (5.103) we integrate only on Ω_{ℓ_1}, we derive

$$\int_{\Omega_{\ell_1}} aw\varrho^2 \, dx \leq \frac{C}{\ell_1} \int_{\Omega_{\ell_1}} w \, dx. \tag{5.104}$$

Using now (5.78), (5.89) we derive

$$\int_{\Omega_{\ell_1/2}} (u_\ell - u_\infty)^+ \, dx \leq \frac{C}{\ell_1} \int_{\Omega_{\ell_1}} (u_\ell - u_\infty)^+ \, dx \tag{5.105}$$

where C is a constant independent of $\ell_1 \geq 1$. Since $-u_\ell, -u_\infty$ are functions satisfying similar equations we obtain

$$\int_{\Omega_{\ell_1/2}} |u_\ell - u_\infty| \, dx \leq \frac{C}{\ell_1} \int_{\Omega_{\ell_1}} |u_\ell - u_\infty| \, dx. \tag{5.106}$$

Starting with $\ell_1 = \ell/2^{k-1}$ and iterating this formula we obtain for every k

$$\int_{\Omega_{\ell/2^k}} |u_\ell - u_\infty| \, dx \leq \frac{C2^{k-1}}{\ell} \int_{\Omega_{\ell/2^{k-1}}} |u_\ell - u_\infty| \, dx$$

$$\leq \frac{C_k}{\ell^k} \int_{\Omega_\ell} |u_\ell - u_\infty| \, dx. \tag{5.107}$$

Using now the Cauchy–Schwarz inequality,

$$\int_{\Omega_{\ell/2^k}} |u_\ell - u_\infty| \, dx \leq \frac{C}{\ell^k} \left\{ \int_{\Omega_\ell} |u_\ell| \, dx + \int_{\Omega_\ell} |u_\infty| \, dx \right\}$$

$$\leq \frac{C}{\ell^k} \left\{ \left(\int_{\Omega_\ell} |u_\ell|^2 \, dx \right)^{\frac{1}{2}} |\Omega_\ell|^{\frac{1}{2}} + (2\ell)^p \int_\omega |u_\infty| \, dx_2 \right\}, \tag{5.108}$$

with C independent of ℓ.

We have – see (5.87), (3.28) for $a = 0$

$$|u_\ell|_{2,\Omega_\ell} \leq C||\nabla u_\ell||_{2,\Omega_\ell} \leq C\ell^{\frac{p}{2}}$$

where C is also independent of ℓ. From (5.108),

$$\int_{\Omega_{\ell/2^k}} |u_\ell - u_\infty| \, dx \leq \frac{C}{\ell^k} \ell^p = \frac{C}{\ell^{k-p}}$$

where C is independent of ℓ.

Choosing then $k - p \geq r$, $\ell/2^k \geq \ell_0$ we obtain (5.86). This completes the proof of the theorem. $\qquad\square$

Having established the convergence in L^1-norm, we now have the question of convergence in H^1-norm. This can be established and under the assumptions of Theorem 5.7 we have

THEOREM 5.9. *Suppose that we are under the conditions of Theorem 5.7 with in addition*

$$\beta_j(x_2, 0) = 0 \quad \forall j = p+1, \ldots, n, \quad f \in L^2(\omega). \tag{5.109}$$

Moreover we suppose that (5.81) holds for all the coefficients a_{ij}. Then if u_ℓ, u_∞ are solutions to (5.84), (5.83) it holds that for any $\ell_0 > 0$

$$u_\ell \to u_\infty \quad in \ H^1(\Omega_{\ell_0}). \tag{5.110}$$

PROOF. Note that assuming all the a_{ij} satisfy (5.81) insures the uniqueness of u_ℓ – see Theorem 4.5. We divide the proof into different steps. First we are looking for

• *pointwise estimates of u_ℓ*

Let us denote by f^+, f^- the positive and negative parts of f. Due to Theorem 4.5 there exist unique functions u_∞^+, u_∞^- solutions to

$$\begin{cases} u_\infty^+ \in H_0^1(\omega), \\ \displaystyle\int_\omega A_{22}(\mathrm{x}_2, u_\infty^+)\nabla_{\mathrm{x}_2} u_\infty^+ \nabla_{\mathrm{x}_2} v + \beta_j(\mathrm{x}_2, u_\infty^+)\partial_{x_j} v + a u_\infty^+ v \, d\mathrm{x}_2 \\ \displaystyle\quad = \int_\omega f^+ v \, d\mathrm{x}_2 \quad \forall v \in H_0^1(\omega), \end{cases} \tag{5.111}$$

$$\begin{cases} u_\infty^- \in H_0^1(\omega), \\ \displaystyle\int_\omega A_{22}(\mathrm{x}_2, u_\infty^-)\nabla_{\mathrm{x}_2} u_\infty^- \nabla_{\mathrm{x}_2} v + \beta_j(\mathrm{x}_2, u_\infty^-)\partial_{x_j} v + a u_\infty^- v \, d\mathrm{x}_2 \\ \displaystyle\quad = \int_\omega -f^- v \, d\mathrm{x}_2 \quad \forall v \in H_0^1(\omega). \end{cases} \tag{5.112}$$

For $f = 0$, due to (5.109) it is easy to see that the solution to (5.83) is $u_\infty = 0$. Due to $-f^- \leq 0 \leq f^+$ and Remark 4.1 we have

$$u_\infty^- \leq 0 \leq u_\infty^+ \quad \text{in } \omega. \tag{5.113}$$

Next taking as test function in (5.111), (5.112), $v = v(\mathrm{x}_1, \cdot)$ where $v \in H_0^1(\Omega_\ell)$ and integrating the equations obtained on $(-\ell, \ell)^p$ we see easily that u_∞^+, u_∞^- satisfy

$$\int_{\Omega_\ell} A\nabla u_\infty^\pm \nabla v + \beta_j(\mathrm{x}_2, u_\infty^\pm)\partial_{x_j} v + a u_\infty^\pm v \, dx = \int_{\Omega_\ell} \pm f^\pm v \, dx \quad \forall v \in H_0^1(\Omega_\ell).$$

In particular choosing $v \geq 0$ we have

$$\int_{\Omega_\ell} A\nabla u_\infty^+ \nabla + \beta_j(\mathrm{x}_2, u_\infty^+)\partial_{x_j} v + a u_\infty^+ v \, dx$$

$$= \int_{\Omega_\ell} f^+ v \, dx \geq \int_{\Omega_\ell} fv \tag{5.114}$$

$$= \int_{\Omega_\ell} A\nabla u_\ell \nabla v + \beta_j(\mathrm{x}_2, u_\ell)\partial_{x_j} v + a u_\ell v \, dx \quad \forall v \in H_0^1(\Omega_\ell), \quad v \geq 0.$$

This and

$$u_\infty^+ \geq u_\ell \quad \text{on} \quad \partial\Omega_\ell, \tag{5.115}$$

where $\partial\Omega_\ell$ is the boundary of Ω_ℓ, imply (argue as in Theorem 4.5)

$$u_\ell \leq u_\infty^+ \quad \text{in} \quad \Omega_\ell.$$

Arguing similarly for u_∞^- we have obtained

$$u_\infty^- \leq u_\ell \leq u_\infty^+ \quad \text{in} \quad \Omega_\ell. \tag{5.116}$$

Then we try to establish

- *local estimates for u_ℓ*

For that we consider $\varrho = \varrho(\mathrm{x}_1)$ a smooth function such that

$$0 \leq \varrho \leq 1, \quad \varrho = 1 \text{ on } (-\ell_0, \ell_0)^p, \quad \varrho = 0 \text{ on } \mathbb{R}^p \setminus (-\ell_0 - 1, \ell_0 + 1)^p. \tag{5.117}$$

Using $v = u_\ell \varrho^2$ in (5.84) leads to

$$\int_{\Omega_{\ell_0+1}} A\nabla u_\ell \nabla(u_\ell \varrho^2) + \beta_j(\mathrm{x}_2, u_\ell)\partial_{x_j}(u_\ell \varrho^2) + au_\ell^2 \varrho^2 \, dx$$

$$= \langle f, u_\ell \varrho^2 \rangle = \int_{\Omega_{\ell_0+1}} fu_\ell \varrho^2.$$

Performing the derivation in the first integral we get

$$\int_{\Omega_{\ell_0+1}} A\nabla u_\ell \nabla u_\ell \varrho^2 + au_\ell^2 \varrho^2 \, dx$$

$$= \int_{\Omega_{\ell_0+1}} fu_\ell \varrho^2 \, dx - \int_{\Omega_{\ell_0+1}} \beta_j(\mathrm{x}_2, u_\ell)\partial_{x_j} u_\ell \varrho^2 \, dx \qquad (5.118)$$

$$- 2\int_{\Omega_{\ell_0+1}} A\nabla u_\ell \nabla \varrho u_\ell \varrho \, dx.$$

Using on the right-hand side of (5.118) the Cauchy–Schwarz inequality and (5.117) we obtain, thanks also to the ellipticity of A,

$$\int_{\Omega_{\ell_0+1}} (\lambda|\nabla u_\ell|^2 + au_\ell^2)\varrho^2 \, dx \le |f|_{2,\Omega_{\ell_0+1}}|u_\ell \varrho|_{2,\Omega_{\ell_0+1}} + \Lambda|\Omega_{\ell_0+1}|^{1/2}||\nabla u_\ell \varrho||_{2,\Omega_{\ell_0+1}}$$

$$+ C||\nabla u_\ell \varrho||_{2,\Omega_{\ell_0+1}}|u_\ell|_{2,\Omega_{\ell_0+1}}.$$

It follows easily that for some constant C independent of ℓ it holds that

$$\int_{\Omega_{\ell_0+1}} (|\nabla u_\ell|^2 + u_\ell^2)\varrho^2 \, dx \le C(1 + |u_\ell|_{2,\Omega_{\ell_0+1}}) \le C(\ell_0)$$

(by (5.116) – note that u_∞^+, u_∞^- are independent of ℓ). Thus by (5.117) we have obtained

$$\int_{\Omega_{\ell_0}} |\nabla u_\ell|^2 + u_\ell^2 \, dx \le C(\ell_0) \qquad (5.119)$$

where $C(\ell_0)$ is independent of ℓ. ($C(\ell_0)$ depends not only on ℓ_0 but also on λ, Λ, n, f but certainly not on ℓ).

We next pass to the final step of the proof which is

• *convergence of u_ℓ toward u_∞*

From (5.119) it is clear that we can extract a subsequence of ℓ such that – if we denote this subsequence by ℓ_k –

$$u_{\ell_k} \rightharpoonup u \quad \text{in } H^1(\Omega_{\ell_0}). \qquad (5.120)$$

This would imply in particular that

$$u_{\ell_k} \rightharpoonup u \quad \text{in } \mathcal{D}'(\Omega_{\ell_0})$$

i.e., in the distributional sense in Ω_{ℓ_0}. But due to Theorem 5.7 we have also

$$u_{\ell_k} \to u_\infty \quad \text{in } \mathcal{D}'(\Omega_{\ell_0}),$$

and by the uniqueness of the limit in $\mathcal{D}'$, $u = u_\infty$. The weak limit being the same for any subsequence we have in fact

$$u_\ell \rightharpoonup u_\infty \quad \text{in } H^1(\Omega_{\ell_0}) \quad \forall \ell_0 > 0. \tag{5.121}$$

Next we pass to the strong convergence. We consider again the function ϱ of (5.117). Using in (5.83), (5.84) the test function

$$(u_\ell - u_\infty)\varrho^2$$

we derive easily that

$$\begin{aligned}
\int_{\Omega_{\ell_0+1}} & A\nabla u_\ell \nabla \{(u_\ell - u_\infty)\varrho^2\} + \beta_j(\mathrm{x}_2, u_\ell)\partial_{x_j}\{(u_\ell - u_\infty)\varrho^2\} \\
& + a u_\ell(u_\ell - u_\infty)\varrho^2 \, dx \\
= \int_{\Omega_{\ell_0+1}} & A_{22}(\mathrm{x}_2, u_\infty)\nabla_{\mathrm{X}_2} u_\infty \nabla_{\mathrm{X}_2}(u_\ell - u_\infty)\varrho^2 \\
& + \beta_j(\mathrm{x}_2, u_\infty)\partial_{x_j}(u_\ell - u_\infty)\varrho^2 + a u_\infty(u_\ell - u_\infty)\varrho^2 \, dx.
\end{aligned} \tag{5.122}$$

We remark first that

$$\begin{aligned}
\int_{\Omega_{\ell_0+1}} & A(x, u_\infty)\nabla u_\infty \nabla \{(u_\ell - u_\infty)\varrho^2\} \\
= \int_{\Omega_{\ell_0+1}} & A_{12}(\mathrm{x}_2, u_\infty)\nabla_{\mathrm{X}_2} u_\infty \nabla_{\mathrm{X}_1}\{(u_\ell - u_\infty)\varrho^2\} \, dx \\
& + \int_{\Omega_{\ell_0+1}} A_{22}(\mathrm{x}_2, u_\infty)\nabla_{\mathrm{X}_2} u_\infty \nabla_{\mathrm{X}_2}(u_\ell - u_\infty)\varrho^2 \, dx \\
= \int_{\Omega_{\ell_0+1}} & A_{22}(\mathrm{x}_2, u_\infty)\nabla_{\mathrm{X}_2} u_\infty \nabla_{\mathrm{X}_2}(u_\ell - u_\infty)\varrho^2 \, dx.
\end{aligned} \tag{5.123}$$

Thus from (5.122) we derive then easily

$$\begin{aligned}
\int_{\Omega_{\ell_0+1}} & \{A\nabla(u_\ell - u_\infty)\nabla(u_\ell - u_\infty) + a(u_\ell - u_\infty)^2\}\varrho^2 \, dx \\
= - \int_{\Omega_{\ell_0+1}} & A\nabla u_\ell \nabla \varrho^2(u_\ell - u_\infty) \, dx \\
& - \int_{\Omega_{\ell_0+1}} \{\beta_j(\mathrm{x}_2, u_\ell) - \beta_j(\mathrm{x}_2, u_\infty)\}\partial_{x_j}(u_\ell - u_\infty)\varrho^2 \, dx \\
= - \sum_{\substack{i=1,\ldots,p \\ j=1,\ldots,n}} & \int_{\Omega_{\ell_0+1}} a_{ij}(x, u_\ell)\partial_{x_j} u_\ell \partial_{x_i} \varrho^2(u_\ell - u_\infty) \, dx \\
& - \int_{\Omega_{\ell_0+1}} \{\beta_j(\mathrm{x}_2, u_\ell) - \beta_j(\mathrm{x}_2, u_\infty)\}\partial_{x_j}(u_\ell - u_\infty)\varrho^2 \, dx.
\end{aligned} \tag{5.124}$$

We have for some constant C, $\forall i = 1, \ldots, p$, $j = 1, \ldots, n$,

$$|a_{ij}(x, u_\ell)\partial_{x_i}\varrho^2(u_\ell - u_\infty)|_{2,\Omega_{\ell_0+1}} \leq C|u_\ell - u_\infty|_{2,\Omega_{\ell_0+1}} \to 0. \tag{5.125}$$

(Indeed the canonical embedding from $H^1(\Omega_{\ell_0+1})$ into $L^2(\Omega_{\ell_0+1})$ being compact this last convergence follows from (5.121) – note that the limit there is unique.) Moreover we claim that

$$|\{\beta_j(x_2, u_\ell) - \beta_j(x_2, u_\infty)\}\varrho^2|_{2,\Omega_{\ell_0+1}} \to 0. \tag{5.126}$$

If not, for some ε and some subsequence ℓ_k we would have

$$|\{\beta_j(x_2, u_{\ell_k}) - \beta_j(x_2, u_\infty)\}\varrho^2|_{2,\Omega_{\ell_0+1}} \geq \varepsilon > 0. \tag{5.127}$$

But then, since $u_\ell \to u_\infty$ in $L^2(\Omega_{\ell_0+1})$, up to a subsequence we have,

$$u_{\ell_k} \to u_\infty \quad \text{a.e. in } \Omega_{\ell_0+1}.$$

By the Lebesgue theorem – recall that β_j is bounded – it holds that

$$|\beta_j(x_2, u_{\ell_k})\varrho^2 - \beta_j(x_2, u_\infty)\varrho^2|_{2,\Omega_{\ell_0+1}} \to 0$$

which contradicts (5.127). Combining (5.124)–(5.126) we can easily pass to the limit in (5.124) so that

$$\lim_{\ell \to +\infty} \int_{\Omega_{\ell_0+1}} \{A\nabla(u_\ell - u_\infty)\nabla(u_\ell - u_\infty) + a(u_\ell - u_\infty)^2\}\varrho^2 \, dx = 0. \tag{5.128}$$

From (5.117) we derive easily that

$$\min(\lambda, \alpha) \int_{\Omega_{\ell_0}} |\nabla(u_\ell - u_\infty)|^2 + (u_\ell - u_\infty)^2 \, dx$$

$$\leq \int_{\Omega_{\ell_0}} \lambda|\nabla(u_\ell - u_\infty)|^2 + \alpha(u_\ell - u_\infty)^2 \, dx$$

$$\leq \int_{\Omega_{\ell_0}} A\nabla(u_\ell - u_\infty)\nabla(u_\ell - u_\infty) + a(u_\ell - u_\infty) \, dx$$

$$\leq \int_{\Omega_{\ell_0+1}} \{A\nabla(u_\ell - u_\infty)\nabla(u_\ell - u_\infty) + a(u_\ell - u_\infty)^2\}\varrho^2 \, dx$$

and the result – i.e., (5.110) – follows from (5.128). □

Open problems

1. Under the assumptions of Theorem 5.1, is it possible to find estimates of $u_\ell - u_\infty$ in $H^2(\Omega_{\ell_0})$?

2. Under the assumption of Theorem 5.1, could we estimate directly

$$|u_\ell - u_\infty|_{\infty,\Omega_{\ell_0}}?$$

3. When is it possible in Theorem 5.3 to obtain an exponential rate of convergence in ℓ?

4. If one considers the solution to

$$\begin{cases} \displaystyle\int_{\Omega_\ell} A^\ell \nabla u_\ell \nabla(v - u_\ell) + a^\ell u_\ell(v - u_\ell)\,dx \\ \qquad\qquad\qquad \geq \langle f^\ell, v - u_\ell \rangle \quad \forall v \in K^\ell, \\ u_\ell \in K^\ell, \end{cases}$$

find the minimal assumptions

$$A_{22}^\ell \to A_{22}(\mathrm{X}_2), \quad a^\ell \to a(\mathrm{X}_2), \quad f^\ell \to f(\mathrm{X}_2)$$

in order for u_ℓ to converge toward u_∞ the solution to

$$\begin{cases} \displaystyle\int_\omega A_{22}\nabla_{\mathrm{X}_2} u_\infty \nabla_{\mathrm{X}_2}(v - u_\infty) + au_\infty v - u_\infty\,d\mathrm{X}_2 \\ \qquad\qquad\qquad \geq \langle f, v - u_\infty \rangle \quad \forall v \in K, \\ u_\infty \in K. \end{cases}$$

5. The key points in Theorem 5.3 are (5.31), (5.33). Would it be possible to obtain convergence of u_ℓ toward u_∞ even if these assumptions are not fulfilled? In other words what is the minimal subset K^ℓ to consider in order to have convergence? For instance in (5.45) can we replace K^ℓ by

$$\overline{K}^\ell = \{\, v \in H_0^1(\Omega_\ell) \mid |\nabla v(x)| \leq 1 \text{ a.e. } x \in \Omega_\ell \,\}?$$

6. It would be interesting to extend Theorem 5.7 in the case where u_∞ is solution to

$$\begin{cases} u_\infty \in H_0^1(\omega; \partial_0\omega), \\ \displaystyle\int_\omega A_{22}(\mathrm{X}_2, u_\infty)\nabla_{\mathrm{X}_2} u_\infty \nabla_{\mathrm{X}_2} v + \beta_j(\mathrm{X}_2, u_\infty)\partial_{x_j} v \\ \qquad + a(\mathrm{X}_2)u_\infty v\,d\mathrm{X}_2 = \langle f, v \rangle \quad \forall v \in H_0^1(\omega; \partial_0\omega) \end{cases}$$

and u_ℓ solution to

$$\begin{cases} u_\ell \in H_0^1(\Omega_\ell; \Gamma_0), \\ \displaystyle\int_{\Omega_\ell} A\nabla u_\ell \nabla v + \beta_j(\mathrm{X}_2, u_\ell)\partial_{x_j} v \\ \qquad + a(\mathrm{X}_2)u_\ell v\,dx = \langle f, v \rangle \quad \forall v \in H_0^1(\Omega_\ell; \Gamma_0), \end{cases}$$

where $\Gamma_0 = \partial(-\ell, \ell)^p \times \omega \cup (-\ell, \ell) \times \partial_0\omega$, f is an element of the dual of $H_0^1(\omega; \partial_0\omega)$.

7. In Theorem 5.9 could it be possible to estimate the rate of convergence of u_ℓ toward u_∞? In other words is it possible for some constant C independent of ℓ to obtain estimates of the type

$$|u_\ell - u_\infty|_{H^1(\Omega_{\ell_0})} \leq \frac{C}{\ell^r}, \qquad |u_\ell - u_\infty|_{H^1(\Omega_{\ell_0})} \leq Ce^{-\alpha\ell}?$$

Is it also possible to prove Theorem 5.9 under the conditions of Theorem 5.7?

Chapter 6

Elliptic Systems

In this chapter we are going to develop an elementary theory for elliptic systems.
Our main application will be devoted to the system of elasticity.

6.1. Some inequalities

The unknown of the problem is now no more a function u but a vector

$$u = (u^1, \ldots, u^N). \tag{6.1}$$

Consider for instance Ω a bounded open subset of $\mathbb{R}^n$. Given functions or coefficients $A^{ij}_{\alpha\beta} = A^{ij}_{\alpha\beta}(x)$, $i, j = 1, \ldots, n$, $\alpha, \beta = 1, \ldots, N$ an elliptic system is the conjunction of N equations in Ω

$$-\partial_{x_i}(A^{ij}_{\alpha\beta}(x)\partial_{x_j}u^\beta) = f_\alpha \quad \forall \alpha = 1, \ldots, N. \tag{6.2}$$

In the above system we made the convention of summation of repeated indices i.e.,
two indices appearing at different levels are summed, $f = (f_1, \ldots, f_N)$ will be a
function or a distribution in Ω. Note that in this introduction we do not consider
lower order terms – i.e., terms without derivatives in the system (6.2). In the case
of systems the ellipticity condition can take several forms i.e., the quadratic form
defined by the $A^{ij}_{\alpha\beta}$ can be positive definite for every matrix or positive definite
only for the rank-one matrices or more generally for some subclass of matrices.
This is the difference between the *Legendre condition* and the *Legendre–Hadamard
condition*. The first one is the existence of a positive λ such that

$$\text{(L)} \qquad A^{ij}_{\alpha\beta}(x)\xi^\alpha_i \xi^\beta_j \geq \lambda|\xi|^2 \quad \forall \xi \in \mathbb{R}^{n \times N}, \quad \text{a.e. } x \in \Omega. \tag{6.3}$$

($|\xi|$ denotes the euclidean norm of the matrix ξ i.e. $|\xi|^2 = \sum_{i,\alpha}(\xi^\alpha_i)^2$.) The
Legendre–Hadamard condition is the existence of a positive λ such that

$$\text{(L–H)} \quad A^{ij}_{\alpha\beta}(x)\varphi_i\varphi_j\eta^\alpha\eta^\beta \geq \lambda|\varphi|^2|\eta|^2 \quad \forall \varphi \in \mathbb{R}^n, \ \forall \eta \in \mathbb{R}^N, \text{ a.e. } x \in \Omega. \tag{6.4}$$

($|\varphi|$, $|\eta|$ denote the usual euclidean norms in $\mathbb{R}^n$ and $\mathbb{R}^N$ respectively, recall our
summation convention.) In the case where $N = 1$ it is clear that both (L) and
(L–H) reduce to the usual ellipticity condition. In the case where $N > 1$ then,
clearly, (L) implies (L–H) since (6.4) is just (6.3) reduced to the rank-one matrices
$(\xi^\alpha_i) = (\varphi_i\eta^\alpha)$. Of course, the two conditions are not equivalent as shown in the
proposition below.

PROPOSITION 6.1. *There exist systems of coefficients $A^{ij}_{\alpha\beta}$ such that (L–H)
holds but (L) fails.*

PROOF. For simplicity we consider the case $n = N = 2$. If (ξ_i^α) is a two-by-two matrix, define $A_{\alpha\beta}^{ij}$ $(i, j, \alpha, \beta = 1, 2)$ as coefficients such that

$$A_{\alpha\beta}^{ij}\xi_i^\alpha\xi_j^\beta = |\xi|^2 + k \det \xi \tag{6.5}$$

where $|\xi|^2 = \sum_{i,\alpha}(\xi_i^\alpha)^2$, $\det \xi$ is the determinant of the matrix ξ and k some constant. Since the right-hand side of (6.5) is a quadratic form we can clearly find such coefficients $A_{\alpha\beta}^{ij}$. If $\xi_i^\alpha = (\varphi_i\eta^\alpha)$ we have of course

$$\det \xi = 0$$

and (6.5) becomes

$$A_{\alpha\beta}^{ij}\varphi_i\varphi_j\eta^\alpha\eta^\beta = |\varphi|^2|\eta|^2 \tag{6.6}$$

so that (L–H) holds whatever k is. Now, if we choose for ξ a nonsingular matrix – i.e., such that

$$\det \xi \neq 0,$$

then provided we choose k of the opposite sign of $\det \xi$ large enough, it holds that

$$A_{\alpha\beta}^{ij}\xi_i^\alpha\xi_j^\beta = |\xi|^2 + k \det \xi < 0 \tag{6.7}$$

which makes (L) impossible and completes the proof. □

REMARK 6.1. Condition (L) is very convenient in order to get coerciveness since it implies for every $u \in \mathbb{H}^1(\Omega)$

$$\int_\Omega A_{\alpha\beta}^{ij}\partial_{x_j}u^\beta\partial_{x_i}u^\alpha \, dx \geq \lambda \int_\Omega |\nabla u|^2 \, dx. \tag{6.8}$$

In the above formula ∇u is the matrix with entries $(\partial_{x_i}u^\alpha)$ – i.e. the Jacobian matrix of u, $|\nabla u|$ is the euclidean norm of ∇u, $\mathbb{H}^1(\Omega)$ denotes the space $(H^1(\Omega))^N$.

In the case where only (L–H) holds we can show

PROPOSITION 6.2. Let $A_{\alpha\beta}^{ij}$ be constants satisfying the Legendre–Hadamard condition. Then it holds that

$$\int_\Omega A_{\alpha\beta}^{ij}\partial_{x_j}u^\beta\partial_{x_i}u^\alpha \, dx \geq \lambda \int_\Omega |\nabla u|^2 \, dx \quad \forall u \in \mathbb{H}_0^1(\Omega). \tag{6.9}$$

(As above ∇u denotes the matrix with entries $(\partial_{x_i}u^\alpha)$, $\mathbb{H}_0^1(\Omega) = (H_0^1(\Omega))^N$.)

PROOF. By density of $\mathcal{D}(\Omega)$ in $H_0^1(\Omega)$ it is enough to prove (6.9) in the case where $u \in \mathcal{D}(\Omega)^N$. In the computation below we assume that all the functions have been extended by 0 outside of Ω. Since we suppose all the functions we are dealing with to be real valued, it holds that

$$I \stackrel{\text{Def}}{=} \int_\Omega A_{\alpha\beta}^{ij}\partial_{x_j}u^\beta\partial_{x_i}u^\alpha \, dx$$
$$= \frac{1}{2}\left\{\int_{\mathbb{R}^n} \overline{A_{\alpha\beta}^{ij}\partial_{x_j}u^\beta}\partial_{x_i}u^\alpha \, dx + \int_{\mathbb{R}^n} A_{\alpha\beta}^{ij}\partial_{x_j}u^\beta\overline{\partial_{x_i}u^\alpha} \, dx\right\}. \tag{6.10}$$

($\bar{z}$ denotes the conjugate of the complex number z.) We denote by $\hat{f}$ the Fourier transform of the function f – i.e., we set

$$\hat{f}(y) = \int_{\mathbb{R}^n} e^{-2\pi i y x} f(x)\, dx \tag{6.11}$$

the above formula being well defined for $f \in \mathcal{D}(\Omega)$. Going back to (6.10) we get by the Plancherel formula (see for instance [33] for details)

$$I = \frac{1}{2}\left\{ \int_{\mathbb{R}^n} \overline{A^{ij}_{\alpha\beta}\widehat{\partial_{x_j}u^\beta}}\,\widehat{\partial_{x_i}u^\alpha}\, dy + \int_{\mathbb{R}^n} A^{ij}_{\alpha\beta}\widehat{\partial_{x_j}u^\beta}\,\overline{\widehat{\partial_{x_i}u^\alpha}}\, dy \right\}. \tag{6.12}$$

Since the $A^{ij}_{\alpha\beta}$ are real constants, we get after some easy manipulations,

$$I = \frac{1}{2}\left\{ \int_{\mathbb{R}^n} A^{ij}_{\alpha\beta}\overline{\widehat{\partial_{x_j}u^\beta}}\,\widehat{\partial_{x_i}u^\alpha}\, dy + \int_{\mathbb{R}^n} A^{ij}_{\alpha\beta}\widehat{\partial_{x_j}u^\beta}\,\overline{\widehat{\partial_{x_i}u^\alpha}}\, dy \right\} \tag{6.13}$$

If it holds that

$$\widehat{u^\alpha} = a^\alpha + i b^\alpha,$$

then we have

$$\mathrm{Re}(\overline{\widehat{u^\alpha}\widehat{u^\beta}}) = \frac{1}{2}(\overline{\widehat{u^\alpha}\widehat{u^\beta}} + \widehat{u^\alpha}\overline{\widehat{u^\beta}}) = a^\alpha a^\beta + b^\alpha b^\beta.$$

Thus from (6.13) we derive using (L–H)

$$\begin{aligned}
I &= \int_{\mathbb{R}^n} (2\pi)^2 A^{ij}_{\alpha\beta} y_j y_i \{a^\alpha a^\beta + b^\alpha b^\beta\}\, dy \\
&\geq \lambda \int_{\mathbb{R}^n} (2\pi)^2 |y|^2 \{|a|^2 + |b|^2\}\, dy \\
&= \lambda \int_{\mathbb{R}^n} (2\pi)^2 |y|^2 |\hat{u}|^2\, dy = \lambda \int_{\mathbb{R}^n} \overline{\widehat{\partial_{x_i}u^\alpha}}\,\widehat{\partial_{x_i}u^\alpha}\, dx
\end{aligned} \tag{6.14}$$

(with the summation convention). Thus we have obtained

$$I \geq \lambda \int_\Omega |\nabla u|^2\, dx. \tag{6.15}$$

This completes the proof of the proposition. □

In the case of variable coefficients we can show

PROPOSITION 6.3 (Gårding inequality). *Let us assume that $A^{ij}_{\alpha\beta} = A^{ij}_{\alpha\beta}(x)$ are continuous functions on $\overline{\Omega}$ satisfying (L–H). Then, there exists λ_1, λ_2 two positive constants such that*

$$\int_\Omega A^{ij}_{\alpha\beta}\partial_{x_j}u^\beta \partial_{x_i}u^\alpha\, dx \geq \lambda_1 \int_\Omega |\nabla u|^2\, dx - \lambda_2 \int_\Omega |u|^2\, dx \quad \forall\, u \in \mathbb{H}^1_0(\Omega). \tag{6.16}$$

(As above $\mathbb{H}^1_0(\Omega) = (H^1_0\Omega))^N$, $|\cdot|$ denotes the euclidean norm of vectors or matrices.)

PROOF. Let us divide the proof into two steps.

Step 1. Let $B_\varepsilon(x_0)$ be a ball of center $x_0 \in \overline{\Omega}$ and radius ε.

Let $u \in (\mathcal{D}(B_\varepsilon(x_0) \cap \Omega))^N$. Then we have

$$\int_\Omega A_{\alpha\beta}^{ij}(x)\partial_{x_j}u^\beta\partial_{x_i}u^\alpha\,dx$$

$$= \int_{B_\varepsilon(x_0)} A_{\alpha\beta}^{ij}(x_0)\partial_{x_j}u^\beta\partial_{x_i}u^\alpha\,dx$$

$$+ \int_{B_\varepsilon(x_0)} (A_{\alpha\beta}^{ij}(x) - A_{\alpha\beta}^{ij}(x_0))\partial_{x_j}u^\beta\partial_{x_i}u^\alpha\,dx.$$

Using (6.9) for the first integral above we obtain

$$\int_\Omega A_{\alpha\beta}^{ij}(x)\partial_{x_j}u^\beta\partial_{x_i}u^\alpha\,dx$$

$$\geq \lambda\int_\Omega |\nabla u|^2\,dx + \int_{B_\varepsilon(x_0)}(A_{\alpha\beta}^{ij}(x) - A_{\alpha\beta}^{ij}(x_0))\partial_{x_j}u^\beta\partial_{x_i}u^\alpha\,dx$$

$$\geq \lambda\int_\Omega |\nabla u|^2\,dx - \sum_{i,j,\alpha,\beta}\int_{B_\varepsilon(x_0)} |A_{\alpha\beta}^{ij}(x) - A_{\alpha\beta}^{ij}(x_0)|\,|\nabla u|^2\,dx \tag{6.17}$$

$$\geq \left\{\lambda - n^2 N^2 \operatorname*{Max}_{i,j,\alpha,\beta}\operatorname*{Sup}_{B_\varepsilon(x_0)} |A_{\alpha\beta}^{ij}(x) - A_{\alpha\beta}^{ij}(x_0)|\right\}\int_\Omega |\nabla u|^2\,dx.$$

Since the $A_{\alpha\beta}^{ij}(x)$ are uniformly continuous in $\overline{\Omega}$ we can find ε small enough such that

$$n^2 N^2 \operatorname*{Max}_{i,j,\alpha,\beta}\operatorname*{Sup}_{B_\varepsilon(x_0)} |A_{\alpha\beta}^{ij}(x) - A_{\alpha\beta}^{ij}(x_0)| \leq \frac{\lambda}{2}$$

and (6.17) becomes

$$\int_\Omega A_{\alpha\beta}^{ij}(x)\partial_{x_j}u^\beta\partial_{x_i}u^\alpha\,dx \geq \frac{\lambda}{2}\int_\Omega |\nabla u|^2\,dx. \tag{6.18}$$

Step 2. We choose ε small enough such that (6.18) holds for every ball $B_\varepsilon(x_0)$ –

i.e., for every $u \in \mathcal{D}(B_\varepsilon(x_0) \cap \Omega)^N$. Then, due to the compactness of $\overline{\Omega}$, we can cover it by a finite number of balls $B_\varepsilon(x_k)$, $k = 1, \ldots, p$. By a well-known theorem (see for instance [34]) we can find functions ϕ_k, $k = 1, \ldots, p$ such that

$$\phi_k \in \mathcal{D}(B_\varepsilon(x_k)), \qquad \sum_{k=1}^p \phi_k^2 = 1. \tag{6.19}$$

Due to the definition of $\mathbb{H}_0^1(\Omega)$ it is clearly enough to prove (6.16) for u such that $u \in \mathcal{D}(\Omega)^N$. For such a u it holds that

$$
\begin{aligned}
I &= \int_\Omega A_{\alpha\beta}^{ij}(x)\partial_{x_j}u^\beta\partial_{x_i}u^\alpha\,dx \\
&= \int_\Omega A_{\alpha\beta}^{ij}(x)\Big(\sum_{k=1}^p \phi_k^2\Big)\partial_{x_j}u^\beta\partial_{x_i}u^\alpha\,dx \\
&= \sum_{k=1}^p\Big\{\int_\Omega A_{\alpha\beta}^{ij}(x)\partial_{x_j}(\phi_k u^\beta)\partial_{x_i}(\phi_k u^\alpha)\,dx - \int_\Omega A_{\alpha\beta}^{ij}(x)\partial_{x_j}\phi_k u^\beta\partial_{x_i}u^\alpha\phi_k \quad (6.20) \\
&\quad - \int_\Omega A_{\alpha\beta}^{ij}(x)\partial_{x_j}u^\beta\phi_k\partial_{x_i}\phi_k u^\alpha\,dx \\
&\quad - \int_\Omega A_{\alpha\beta}^{ij}(x)\partial_{x_j}\phi_k\partial_{x_i}\phi_k u^\beta u^\alpha\,dx\Big\}.
\end{aligned}
$$

Using (6.18) and the fact that the functions $A_{\alpha\beta}^{ij}$, $\partial_{x_\ell}\phi_k$ are bounded we obtain

$$
I \ge \frac{\lambda}{2}\sum_{k=1}^p\int_\Omega |\nabla(\phi_k u)|^2\,dx - C_1\int_\Omega |u|\,|\nabla u|\,dx - C_2\int_\Omega |u|^2\,dx \qquad (6.21)
$$

for some constants C_1, C_2 depending on the $A_{\alpha\beta}^{ij}$ and ϕ_k but independent of u.

Next we have

$$
\begin{aligned}
(\partial_{x_\ell}(\phi_k u^\gamma))^2 &= (\phi_k\partial_{x_\ell}u^\gamma + u^\gamma\partial_{x_\ell}\phi_k)^2 \\
&= \phi_k^2(\partial_{x_\ell}u^\gamma)^2 + 2\phi_k\partial_{x_\ell}\phi_k u^\gamma\partial_{x_\ell}u^\gamma + (u^\gamma)^2(\partial_{x_\ell}\phi_k)^2
\end{aligned}
$$

and from (6.21) we derive – recall that $|\nabla(\phi_k u)|$ denotes the euclidean norm of the matrix $\nabla(\phi_k u)$ – for some other constants C_3, C_4,

$$
\begin{aligned}
I &\ge \frac{\lambda}{2}\sum_{k=1}^p\int_\Omega \phi_k^2|\nabla u|^2\,dx - C_3\int_\Omega |u|\,|\nabla u|\,dx - C_4\int_\Omega |u|^2\,dx \\
&= \frac{\lambda}{2}\int_\Omega |\nabla u|^2\,dx - C_3\int_\Omega |u|\,|\nabla u|\,dx - C_4\int_\Omega |u|^2\,dx.
\end{aligned}
\qquad (6.22)
$$

Next, applying the Young inequality

$$
ab \le \frac{\lambda a^2}{4C_3} + \frac{C_3 b^2}{\lambda}
$$

with $a = |\nabla u|$, $b = |u|$ we obtain – recall the definition of I

$$
\begin{aligned}
I &= \int_\Omega A_{\alpha\beta}^{ij}(x)\partial_{x_j}u^\beta\partial_{x_i}u^\alpha\,dx \\
&\ge \frac{\lambda}{2}\int_\Omega |\nabla u|^2\,dx - \frac{\lambda}{4}\int_\Omega |\nabla u|^2\,dx - \Big(\frac{C_3^2}{\lambda} + C_4\Big)\int_\Omega |u|^2\,dx.
\end{aligned}
$$

This implies (6.16) with $\lambda_1 = \frac{\lambda}{4}$, $\lambda_2 = \frac{C_3^2}{\lambda} + C_4$. This completes the proof of the theorem. $\qquad\square$

As a corollary we have

COROLLARY 6.4. *Let $A_{\alpha\beta}^{ij} = A_{\alpha\beta}^{ij}(x)$ be continuous functions on $\overline{\Omega}$ satisfying the Legendre–Hadamard condition. Let $b_{\alpha\beta}^i$, $c_{\alpha\beta}$ be $L^\infty(\Omega)$-functions. Then, there exists two constants λ_1, λ_2 such that*

$$\int_\Omega A_{\alpha\beta}^{ij}(x)\partial_{x_j}u^\beta\partial_{x_i}u^\alpha + b_{\alpha\beta}^i u^\beta \partial_{x_i}u^\alpha + c_{\alpha\beta}u^\alpha u^\beta \, dx$$

$$\geq \lambda_1 \int_\Omega |\nabla u|^2 \, dx - \lambda_2 \int_\Omega |u|^2 \, dx \quad \forall u \in \mathbb{H}_0^1(\Omega). \tag{6.23}$$

PROOF. By Proposition 6.3 we have for some constants λ_1, λ_2

$$\int_\Omega A_{\alpha\beta}^{ij}(x)\partial_{x_j}u^\beta\partial_{x_i}u^\alpha \, dx \geq \lambda_1 \int_\Omega |\nabla u|^2 \, dx - \lambda_2 \int_\Omega |u|^2 \, dx \quad \forall u \in \mathbb{H}_0^1(\Omega). \tag{6.24}$$

If A denotes the left-hand side of the inequality (6.23), it follows that for some constants C_1, C_2 it holds that

$$A \geq \lambda_1 \int_\Omega |\nabla u|^2 \, dx - \lambda_2 \int_\Omega |u|^2 \, dx$$

$$- C_1 \int_\Omega |u| \, |\nabla u| \, dx - C_2 \int_\Omega |u|^2 \, dx \quad \forall u \in \mathbb{H}_0^1(\Omega).$$

Then, it is enough to apply the Young inequality to conclude as below (6.22). This concludes the proof of the corollary. □

In the theory of elasticity useful tools are the so-called Korn inequalities. Let us consider this. For that and until the end of Section 6.1 we will assume that

$$n = N. \tag{6.25}$$

In the applications that we will investigate later on, these two numbers will be equal to 2 or 3, but the case n can be handled the same way. Let $u = (u^1, \ldots, u^n)$ be a function from Ω – a bounded open subset of $\mathbb{R}^n$ – into $\mathbb{R}^n$. The so-called strain tensor (see for instance ([15], [16], [25])) is defined as

$$\varepsilon_{ij}(u) = \frac{1}{2}(\partial_{x_i}u^j + \partial_{x_j}u^i) \quad \forall i,j = 1, \ldots, n \tag{6.26}$$

i.e., as the symmetric part of the jacobian matrix ∇u. Then, we have

THEOREM 6.5 (First Korn Inequality). *It holds that*

$$\frac{1}{2}\|\nabla u\|_{2,\Omega}^2 \leq \sum_{i,j=1}^{n} |\varepsilon_{ij}(u)|_{2,\Omega}^2 \quad \forall u \in \mathbb{H}_0^1(\Omega) = H_0^1(\Omega)^n. \tag{6.27}$$

PROOF. It is enough to establish (6.27) for $u \in (\mathcal{D}(\Omega))^n$. For such a u we have

$$|\varepsilon_{ij}(u)|_{2,\Omega}^2 = \frac{1}{4}\int_\Omega (\partial_{x_i}u^j + \partial_{x_j}u^i)^2 \, dx$$

$$= \frac{1}{4}\int_\Omega (\partial_{x_i}u^j)^2 + 2\partial_{x_i}u^j\partial_{x_j}u^i + (\partial_{x_j}u^i)^2 \, dx. \tag{6.28}$$

Integrating by parts – since u is smooth with compact support – we have also

$$\int_\Omega \partial_{x_i} u^j \, \partial_{x_j} u^i \, dx = \int_\Omega \partial_{x_j} u^j \partial_{x_i} u^i \, dx.$$

Taking into account this equality we get by summing up (6.28) in $i, j = 1, \ldots, n$

$$\sum_{i,j=1}^n |\varepsilon_{ij}(u)|_{2,\Omega}^2 = \frac{1}{2} \sum_{i,j=1}^n |\partial_{x_i} u^j|_{2,\Omega}^2 + \frac{1}{2} |\operatorname{div} u|_{2,\Omega}^2 \qquad (6.29)$$

where we have set

$$\operatorname{div} u = \partial_{x_i} u^i \left(= \sum_{i=1}^n \partial_{x_i} u^i \right); \qquad (6.30)$$

(6.27) follows then easily. □

We now turn to the second Korn inequality which is a deeper result. For that we follow the approach of [26].

Let us denote by $\mathbb{R}_+^n$ and $\mathbb{R}_-^n$ the upper and lower half space. In other words

$$\begin{aligned}
\mathbb{R}_+^n &= \{ x = (x_1, \ldots, x_n) \in \mathbb{R}^n \mid x_n > 0 \}, \\
\mathbb{R}_-^n &= \{ x = (x_1, \ldots, x_n) \in \mathbb{R}^n \mid x_n < 0 \}.
\end{aligned} \qquad (6.31)$$

Let us first establish the following lemma.

LEMMA 6.6 (Second Korn inequality in $\mathbb{R}_+^n$). Let $u \in \mathbb{H}^1(\mathbb{R}_+^n) = (H^1(\mathbb{R}_+^n))^n$, u with compact support in $\overline{\mathbb{R}_+^n}$. Then, it holds that

$$\|\nabla u\|_{2,\mathbb{R}_+^n}^2 \leq C \sum_{i,j=1}^n |\varepsilon_{ij}(u)|_{2,\mathbb{R}_+^n}^2 \qquad (6.32)$$

where C is a universal constant independent of u.

PROOF. It is of course enough to show (6.32) for a smooth function u with compact support in $\overline{\mathbb{R}_+^n}$. For such a function u we built an extension $\tilde{u}$ of u to $\mathbb{R}^n$. We proceed as follows. We denote by $x = (x', x_n)$ the points in $\mathbb{R}^n$ where $x' = (x_1, \ldots, x_{n-1})$. Then we set

$$\tilde{u}^i(x) = pu^i(x', -\lambda x_n) + qu^i(x', -\mu x_n), \quad x \in \mathbb{R}_-^n, \quad i = 1, \ldots, n-1, \qquad (6.33)$$

$$\tilde{u}^n(x) = ru^n(x', -\lambda x_n) + su^n(x', -\mu x_n), \quad x \in \mathbb{R}_-^n. \qquad (6.34)$$

$\lambda > 0$, $\mu > 0$, p, q, r, s will be fixed later on. We denote by $\hat{u}$ the function given by

$$\hat{u} = u\chi_{\mathbb{R}_+^n} + \tilde{u}\chi_{\mathbb{R}_-^n} \qquad (6.35)$$

where χ_A is the characteristic function of the set A. Clearly, provided we choose

$$p + q = 1, \qquad r + s = 1 \qquad (6.36)$$

there is no jump between $\tilde{u}$ and u on the hyperplane $x_n = 0$ and it holds that

$$\hat{u} \in \mathbb{H}^1(\mathbb{R}^n), \qquad (6.37)$$

$\hat{u}$ having compact support. Moreover, for $x \in \mathbb{R}^n$ it holds that

$$\varepsilon_{ij}(\hat{u}) = \varepsilon_{ij}(\tilde{u}) = p\varepsilon_{ij}(u)(x', -\lambda x_n) + q\varepsilon_{ij}(u)(x', -\mu x_n) \quad \forall i, j < n, \quad (6.38)$$

$$\varepsilon_{nn}(\hat{u}) = \varepsilon_{nn}(\tilde{u}) = -r\lambda\varepsilon_{nn}(u)(x', -\lambda x_n) - s\mu\varepsilon_{nn}(u)(x', -\mu x_n). \quad (6.39)$$

For $i < n$ it holds that

$$\begin{aligned}
\varepsilon_{in}(\hat{u}) = \varepsilon_{in}(\tilde{u}) &= \frac{1}{2}(\partial_{x_i}\tilde{u}^n + \partial_{x_n}\tilde{u}^i) \\
&= \frac{1}{2}\{r\partial_{x_i}u^n(x', -\lambda x_n) + s\partial_{x_i}u^n(x', -\mu x_n) \\
&\quad - p\lambda\partial_{x_n}u^i(x', -\lambda x_n) - q\mu\partial_{x_n}u^i(x', -\mu x_n)\}.
\end{aligned} \quad (6.40)$$

Choosing

$$r = -p\lambda, \qquad s = -q\mu \quad (6.41)$$

we obtain for $x \in \mathbb{R}^n$

$$\varepsilon_{in}(\hat{u}) = r\varepsilon_{in}(u)(x', -\lambda x_n) + s\varepsilon_{in}(u)(x', -\mu x_n). \quad (6.42)$$

Let us assume that (6.36), (6.41) holds. If we fix $\lambda \neq \mu$ then p, q are determined as solutions to

$$p + q = 1, \qquad \lambda p + \mu q = -1 \quad (6.43)$$

and r, s are given by (6.41). (For instance for $\lambda = 2$, $\mu = 1$ we get $p = -2$, $q = 3$, $r = 4$, $s = -3$.) We have then – thanks to Theorem 6.5

$$\begin{aligned}
\|\nabla u\|_{2,\mathbb{R}^n_+}^2 &\leq \|\nabla \hat{u}\|_{2,\mathbb{R}^n}^2 \leq 2\sum_{i,j}|\varepsilon_{ij}(\hat{u})|_{2,\mathbb{R}^n}^2 \\
&= 2\sum_{i,j}|\varepsilon_{ij}(u)|_{2,\mathbb{R}^n_+}^2 + 2\sum_{i,j}|\varepsilon_{ij}(\tilde{u})|_{2,\mathbb{R}^n_-}^2.
\end{aligned} \quad (6.44)$$

From (6.38) we derive for $i, j \neq n$

$$|\varepsilon_{ij}(\tilde{u})|_{2,\mathbb{R}^n_-} \leq |p|\,|\varepsilon_{ij}(u)(x', -\lambda x_n)|_{2,\mathbb{R}^n_-} + |q|\,|\varepsilon_{ij}(u)(x', -\mu x_n)|_{2,\mathbb{R}^n_-}.$$

Making in the integral the change of variable $z_n = -\lambda x_n$, $z_n = -\mu x_n$ we arrive at

$$|\varepsilon_{ij}(\tilde{u})|_{2,\mathbb{R}^n_-} \leq \left(\frac{|p|}{\sqrt{\lambda}} + \frac{|q|}{\sqrt{\mu}}\right)|\varepsilon_{ij}(u)|_{2,\mathbb{R}^n_+}. \quad (6.45)$$

Arguing the same way for $\varepsilon_{nn}(\tilde{u})$ we derive from (6.39)

$$\begin{aligned}
|\varepsilon_{nn}(\tilde{u})|_{2,\mathbb{R}^n_-} &< (|r|\sqrt{\lambda} + |s|\sqrt{\mu}|\varepsilon_{nn}(u)|_{2,\mathbb{R}^n_+} \\
&< (|p|\lambda\sqrt{\lambda} + |q|\mu\sqrt{\mu})|\varepsilon_{nn}(u)|_{2,\mathbb{R}^n_+}
\end{aligned} \quad (6.46)$$

and from (6.42) for $i \neq n$

$$\begin{aligned}
|\varepsilon_{in}(\tilde{u})|_{2,\mathbb{R}^n_-} &\leq \left(\frac{|r|}{\sqrt{\lambda}} + \frac{|s|}{\sqrt{\mu}}\right)|\varepsilon_{in}(u)|_{2,\mathbb{R}^n_+} \\
&\leq (|p|\sqrt{\lambda} + |q|\sqrt{\mu})|\varepsilon_{in}(u)|_{2,\mathbb{R}^n_+} \quad \text{(by (6.41))}.
\end{aligned} \quad (6.47)$$

Collecting (6.44)–(6.47) we obtain

$$\|\nabla u\|_{2,\mathbb{R}_+^n}^2 \leq 2\{1 + M^2\} \sum_{i,j=1}^n |\varepsilon_{ij}(u)|_{2,\mathbb{R}_+^n}^2$$

where

$$M = \max\left\{ \left(\frac{|p|}{\sqrt{\lambda}} + \frac{|q|}{\sqrt{\mu}}\right), (|p|\lambda\sqrt{\lambda} + |q|\mu\sqrt{\mu}), (|p|\sqrt{\lambda} + |q|\sqrt{\mu}) \right\}.$$

This completes the proof of the theorem. With the values that we have above it holds that

$$M = \max\left\{ \left(\frac{2}{\sqrt{2}} + 3\right), (4\sqrt{2} + 3), (2\sqrt{2} + 3) \right\} = 4\sqrt{2} + 3.$$

$\square$

We turn now to the case of a "smooth" bounded open subset of $\mathbb{R}^n$. We will be in need of the following algebric lemma:

LEMMA 6.7. *Let A, R be $n \times n$ matrices, R orthogonal i.e., $R^t R = {}^t RR = I_n$. If $|\cdot|$ denotes the euclidean norm of matrices it holds that*

$$|A|^2 = \mathrm{tr}(A\,{}^t\!A) = \mathrm{tr}({}^t\!AA) = |AR|^2 = |RA|^2. \tag{6.48}$$

(tr denotes the trace of matrices, t the transposed operator.)

PROOF. Let $A = (a_{ij})$, $i, j = 1, \ldots, n$. The diagonal coefficients of $A^t A$ are given by

$$\sum_{k=1}^n a_{ik} a_{ik} \quad \forall\, i = 1, \ldots, n,$$

and thus it follows that

$$\mathrm{tr}(A\,{}^t\!A) = \mathrm{tr}({}^t\!AA) = \sum_{i=1}^n \sum_{k=1}^n a_{ik}^2 = |A|^2.$$

From that we derive

$$|AR|^2 = \mathrm{tr}(AR\,{}^t\!(AR)) = \mathrm{tr}(AR\,{}^t\!R\,{}^t\!A) = \mathrm{tr}(A\,{}^t\!A) = |A|^2,$$
$$|RA|^2 = \mathrm{tr}({}^t\!(RA)RA) = \mathrm{tr}({}^t\!A\,{}^t\!RRA) = \mathrm{tr}({}^t\!AA) = |A|^2.$$

This completes the proof. $\square$

We now consider the case described by Figure 6.1. More precisely let us consider f a function from $\mathbb{R}^{n-1} \to \mathbb{R}$ such that

$$f(0) = 0, \qquad \nabla f(0) = 0, \qquad f \text{ is of class } C^1. \tag{6.49}$$

Denote by Ω_+ the domain

$$\Omega_+ = \{\, x \in B_R(0) \mid x_n > f(x') \,\} \tag{6.50}$$

where $B_R(0)$ denotes the ball of center 0 and radius R. Then we have

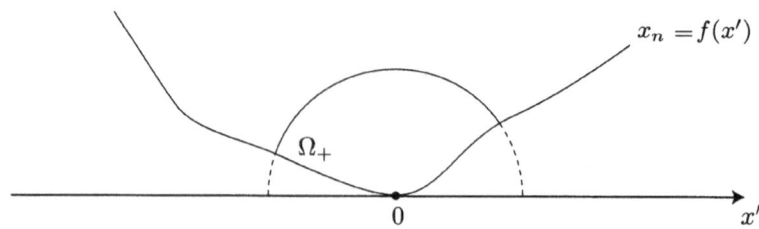

FIGURE 6.1

LEMMA 6.8. *Provided*

$$s = \operatorname*{Sup}_{(x',0)\in B_R(0)} |\nabla f(x')| \tag{6.51}$$

is small enough, there exists a constant C such that

$$||\nabla u||^2_{2,\Omega_+} \le C \sum_{i,j=1}^{n} |\varepsilon_{ij}(u)|^2_{2,\Omega_+} \quad \forall u \in \mathbb{H}^1(\Omega_+),\ u = 0 \text{ on } \partial B_R(0). \tag{6.52}$$

PROOF. Let $u \in \mathbb{H}^1(\Omega_+)$, u vanishing on $\partial B_R(0)$ the boundary of $B_R(0)$. The function

$$v(x', x_n) = u(x', x_n + f(x')) \tag{6.53}$$

is a function of $\mathbb{H}^1(\mathbb{R}^n_+)$ with compact support in $\overline{\mathbb{R}}^n_+$ (we suppose u extended by 0 outside Ω_+). Thus applying Lemma 6.6 we get for some universal constant

$$||\nabla v||_{\mathbb{R}^n_+} \le C \left\{ \sum_{i,j=1}^{n} |\varepsilon_{ij}(v)|^2_{2,\mathbb{R}^n_+} \right\}^{\frac{1}{2}} = C||\varepsilon(v)||_{2,\mathbb{R}^n_+}. \tag{6.54}$$

($\varepsilon(v)$ denotes the matrix with entries $\varepsilon_{ij}(v)$.) Due to (6.53) it holds that

$$\partial_{x_i} v^k(x', x_n) = \partial_{x_i} u^k(x', x_n + f(x')) + \partial_{x_n} u^k(x', x_n + f(x'))\partial_{x_i} f(x') \tag{6.55}$$
$$\forall i = 1, \ldots, n-1,\ \forall k = 1, \ldots, n,$$

$$\partial_{x_n} v^k(x', x_n) = \partial_{x_n} u^k(x', x_n + f(x')) \quad \forall k = 1, \ldots, n. \tag{6.56}$$

From a matrix point of view this can be written as

$$\nabla v(x', x_n) = \nabla u(x', x_n + f(x')) + (\partial_{x_n} u^k(x', x_n + f(x'))\,{}^t(\nabla f(x'), 0). \tag{6.57}$$

(In the above formula we write vectors as columns, t denotes the transposition.) Clearly the Jacobian determinant of the transformation

$$(x', x_n) \to (x', x_n + f(x'))$$

is equal to 1. Thus we obtain easily from (6.57)

$$||\nabla v||_{2,\mathbb{R}^n_+} \ge ||\nabla u||_{2,\Omega_+} - s||\nabla u||_{2,\Omega_+} = (1 - s)||\nabla u||_{2,\Omega_+}. \tag{6.58}$$

It is clear now that the matrix of the $\varepsilon(v)$ is given by

$$\varepsilon(v) = (\varepsilon_{ij}(v)) = \frac{\nabla v + {}^t\nabla v}{2} = \frac{\nabla u + {}^t\nabla u}{2} + \frac{\varrho + {}^t\varrho}{2} = \varepsilon(u) + \frac{1}{2}(\varrho + {}^t\varrho) \quad (6.59)$$

where we have set

$$\varepsilon_{ij}(u) = \varepsilon_{ij}(u)(x', x_n + f(x')) \quad \text{and} \quad \varrho = (\partial_{x_n} u^k(x', x_n + f(x')))^t(\nabla f(x'), 0).$$

If $|\cdot|$ denotes the euclidean norm of matrices it holds that

$$\big|\big|\varepsilon(v)\big|\big|_{2,\mathbb{R}^n_+} \leq \big|\big|\varepsilon(u)\big|\big|_{2,\Omega_+} + s\big|\big|\nabla u\big|\big|_{2,\Omega_+}. \tag{6.60}$$

Recalling (6.54), (6.58) we obtain

$$(1-s)\big|\big|\nabla u\big|\big|_{2,\Omega_+} \leq C\big|\big|\varepsilon(u)\big|\big|_{2,\Omega_+} + Cs\big|\big|\nabla u\big|\big|_{2,\Omega_+}$$

i.e.,

$$\big|\big|\nabla u\big|\big|_{2,\Omega_+} < \frac{C}{(1 - s(1+C))}\big|\big|\varepsilon(u)\big|\big|_{2,\Omega_+}$$

provided $s(1 + C) < 1$. This is nothing but (6.52) and completes the proof of the lemma. $\qquad\square$

We can now give the proof of the second Korn inequality in a bounded open subset of $\mathbb{R}^n$. More precisely we have:

THEOREM 6.9. *Let Ω be a bounded open subset of $\mathbb{R}^n$ with a boundary of class C^1. Then, there exists a constant $C = C(\Omega)$ such that*

$$\big|\big|\nabla u\big|\big|_{2,\Omega} \leq C\{\big|\big|\varepsilon(u)\big|\big|_{2,\Omega} + \big|\big|u\big|\big|_{2,\Omega}\} \quad \forall u \in \mathbb{H}^1(\Omega). \tag{6.61}$$

In the above formula $\varepsilon(u)$ denotes the matrix of $\varepsilon_{ij}(u)$, $|\cdot|$ denotes the Euclidean norm of vectors or matrices.

PROOF. Using a compactness argument we can find q points $\xi_1, \ldots, \xi_q \in \partial\Omega$ the boundary of Ω such that in local coordinates the boundary of Ω is represented by a C^1 function satisfying the assumptions of Lemma 6.8 (see [8]). Moreover, we can cover Ω by the open sets $\Omega_\nu = \Omega \cap B_R(\xi_\nu)$, $\nu = 1, \ldots, q$ and Ω_0 where $\Omega_0 \Subset \Omega$ (see Figure 6.2). Then we consider (φ_ν) a partition of unity associated to this covering i.e., functions φ_ν such that

$$\varphi_\nu \in C^\infty, \qquad \varphi_\nu \text{ has compact support in } \overline{\Omega} \cap B_R(\xi_\nu) \text{ or } \Omega_0,$$

$$0 \leq \varphi_\nu \leq 1, \qquad \sum_{\nu=0}^{q} \varphi_\nu \equiv 1. \tag{6.62}$$

Then, if we set

$$u_\nu = \varphi_\nu u \tag{6.63}$$

it holds that

$$u = \sum_{\nu=0}^{q} u_\nu. \tag{6.64}$$

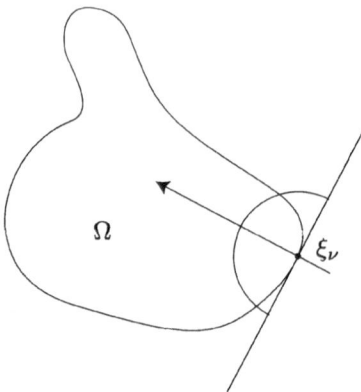

FIGURE 6.2

Thus we have

$$||\nabla u||_{2,\Omega} \le \sum_{\nu=0}^{q}||\nabla u_{\nu}||_{2,\Omega}. \qquad (6.65)$$

Applying Theorem 6.5 we have

$$||\nabla u_0||_{2,\Omega} \le 2||\varepsilon(u_0)||_{2,\Omega}.$$

Moreover, considering u_{ν} we can assume $\xi_{\nu} = 0$ since the partial derivatives are invariant by translation. Then, for some orthogonal transformation R

$$v_{\nu}(x) = u_{\nu}(Rx) \qquad (6.66)$$

is a function to which we can apply Lemma 6.8 and so also is ${}^{t}Rv_{\nu}$. The chain-rule shows that

$$\nabla({}^{t}Rv_{\nu}(x)) = {}^{t}R\nabla u_{\nu}(Rx)R. \qquad (6.67)$$

Thus, from Lemma 6.8 we derive

$$||\nabla({}^{t}Rv_{\nu}(x))||_{2,{}^{t}R\Omega_{\nu}} \le C||\varepsilon({}^{t}Rv_{\nu}(x))||_{2,{}^{t}R\Omega_{\nu}}.$$

(${}^{t}R\Omega_{\nu}$ denotes the set ${}^{t}R(\Omega_{\nu})$.) From (6.67) and Lemma 6.8 we derive

$$||\nabla u_{\nu}(Rx)||_{2,{}^{t}R\Omega_{\nu}} \le C||{}^{t}R\varepsilon(u_{\nu}(Rx))R||_{2,{}^{t}R\Omega_{\nu}}$$
$$= C||\varepsilon(u_{\nu}(Rx))||_{2,{}^{t}R\Omega_{\nu}}.$$

(Recall (6.59)). Doing the change of variable $x \mapsto Rx$ we obtain

$$||\nabla u_{\nu}||_{2,\Omega_{\nu}} \le C||\varepsilon(u_{\nu})||_{2,\Omega_{\nu}} \quad \forall \nu = 1, \ldots, q.$$

Thus from (6.65) we derive

$$\|\nabla u\|_{2,\Omega} \le C \sum_{\nu=0}^{q} \|\varepsilon(u_\nu)\|_{2,\Omega_\nu}. \tag{6.68}$$

Due to (6.63) we have

$$\varepsilon_{ij}(u_\nu) = \varepsilon_{ij}(\varphi_\nu u) = \frac{1}{2}\partial_{x_i}(\varphi_\nu u^j) + \partial_{x_j}(\varphi_\nu u^i)$$

$$= \frac{1}{2}\varphi_\nu \varepsilon_{ij}(u) + \frac{1}{2}\{\partial_{x_i}\varphi_\nu u^j + \partial_{x_j}\varphi_\nu u^i\}. \tag{6.69}$$

Then, it follows easily that

$$\|\nabla u\|_{2,\Omega} \le C\{\|\varepsilon(u)\|_{2,\Omega} + \|u\|_{2,\Omega}\}.$$

This completes the proof of the theorem. □

Let us introduce the following definition:

DEFINITION 6.1 (Rigid Motions). Let Ω be an open connected subset of $\mathbb{R}^n$. A rigid motion is a vector valued distribution – i.e., a vector $u = (u^1, u^2, \ldots, u^n) \in (\mathcal{D}'(\Omega))^n$ such that

$$\varepsilon(u) = 0 \quad \text{in } \mathcal{D}'(\Omega). \tag{6.70}$$

This definition is clarified by the following

THEOREM 6.10. *Let us denote by $\mathcal{R}$ the set of rigid motions in Ω. It holds that*

$$u \in \mathcal{R} \quad \Leftrightarrow \quad u = a + Ax \tag{6.71}$$

where $a \in \mathbb{R}^n$ and A is a skew symmetric matrix of $\mathbb{R}^n$, i.e., a matrix such that

$$^tA = -A \tag{6.72}$$

where ^tA denotes the transposed matrix of A.

PROOF. It is clear when $u = a + Ax$ with $^tA = -A$ that if $A = (a_{ij})$, then

$$\varepsilon_{ij}(u) = \frac{1}{2}\{a_{ij} + a_{ji}\} = 0.$$

Conversely let $u \in (\mathcal{D}'(\Omega))^n$ with

$$\varepsilon_{ij}(u) = \frac{1}{2}\{\partial_{x_j}u^i + \partial_{x_i}u^j\} = 0 \quad \forall i,j = 1,\ldots,n. \tag{6.73}$$

Then for every $i,j,k = 1,\ldots,n$ we have in $\mathcal{D}'(\Omega)$

$$\partial^2_{x_j x_k}u^i = -\partial^2_{x_i x_k}u^j = \partial^2_{x_i x_j}u^k = -\partial^2_{x_k x_j}u^i$$

i.e.,
$$\partial^2_{x_j x_k}u^i = 0 \quad \forall i,j,k = 1,\ldots,n.$$

Since all the second derivatives of u^i vanish, this implies that u^i – for every i – is an affine function. Thus for $a \in \mathbb{R}^n$, $A = (a_{ij})$ an $n \times n$ matrix with real coefficients it holds that

$$u(x) = a + Ax.$$

(6.72) is then an immediate consequence of (6.73). This completes the proof of the theorem.   □

REMARK 6.2. In the case where $n = 3$ the set of rigid motions is the set of transformations of the type

$$u(x) = a + A \wedge x, \qquad a, A \in \mathbb{R}^3. \tag{6.74}$$

We can now show

THEOREM 6.11. *Let Ω be a bounded subset of $\mathbb{R}^n$ with a C^1-boundary. Let V be a closed subspace of $\mathbb{H}^1(\Omega) = (H^1(\Omega))^n$ such that*

$$V \cap \mathcal{R} = \{0\}. \tag{6.75}$$

Then, on V the two norms

$$|u|_{1,2} = \{||\nabla u||_{2,\Omega}^2 + ||u||_{2,\Omega}^2\}^{\frac{1}{2}}, \qquad ||\varepsilon(u)||_{2,\Omega} \tag{6.76}$$

are equivalent.

PROOF. We have obviously

$$||\varepsilon(u)||_{2,\Omega} = \left|\left|\frac{\nabla u + {}^t\nabla u}{2}\right|\right|_{2,\Omega} \le \frac{1}{2}\{||\nabla u||_{2,\Omega} + ||{}^t\nabla u||_{2,\Omega}\}$$
$$\le ||\nabla u||_{2,\Omega} \le |u|_{1,2} \quad \forall u \in \mathbb{H}^1(\Omega). \tag{6.77}$$

Next, suppose that the inequality

$$|u|_{1,2} \le C||\varepsilon(u)||_{2,\Omega} \tag{6.78}$$

does not hold. Then, for every n there exists an element $u_n \in V$ such that

$$|u_n|_{1,2} \ge n||\varepsilon(u_n)||_{2,\Omega}. \tag{6.79}$$

Considering $v_n = u_n/|u_n|_{1,2}$ instead of u_n we can assume that

$$|u_n|_{1,2} = 1. \tag{6.80}$$

Since u_n is a bounded sequence of $\mathbb{H}^1(\Omega)$ we can assume – perhaps only for a subsequence – that there exists $u \in V$ such that

$$u_n \rightharpoonup u \quad \text{in } \mathbb{H}^1(\Omega), \tag{6.81}$$

$$u_n \to u \quad \text{in } \mathbb{L}^2(\Omega). \tag{6.82}$$

From (6.81) we derive in particular that for every $i, j = 1, \ldots, n$

$$\varepsilon_{ij}(u_n) \to \varepsilon_{ij}(u) \quad \text{in } \mathcal{D}'(\Omega). \tag{6.83}$$

But from (6.80), (6.79) we get

$$||\varepsilon(u_n)||_{2,\Omega} \le \frac{1}{n} \tag{6.84}$$

and thus for $i, j = 1, \ldots, n$

$$\varepsilon_{ij}(u_n) \to 0 \text{ in } L^2(\Omega) \text{ and thus also in } \mathcal{D}'(\Omega). \tag{6.85}$$

It results then that
$$u \in \mathcal{R}$$
(see Theorem 6.10). Since $u \in V$ we get by (6.75)
$$u = 0.$$
But then – due to the uniqueness of the possible limit – we derive from (6.82) that
$$u_n \to 0 \quad \text{in } \mathbb{L}^2(\Omega). \tag{6.86}$$
Combining (6.84), (6.86), (6.61) we derive
$$||\nabla u_n||_{2,\Omega} \to 0.$$
Thus – see (6.86) – $|u_n|_{1,2} \to 0$ which contradicts (6.80). This completes the proof of the theorem.   □

As a corollary we have

COROLLARY 6.12. *Let Ω be a bounded domain of $\mathbb{R}^n$ with a C^1 boundary. Let Γ_0 be a subset of $\partial\Omega$ the boundary of Ω with*
$$|\Gamma_0| > 0 \tag{6.87}$$
i.e., assume Γ_0 of positive measure. Then on $\mathbb{H}_0^1(\Omega; \Gamma_0) = H_0^1(\Omega; \Gamma_0)^N$ the norms
$$||\nabla u||_{2,\Omega}, \qquad ||\varepsilon(u)||_{2,\Omega} \tag{6.88}$$
are equivalent.

PROOF. We claim that
$$\mathbb{H}_0^1(\Omega; \Gamma_0) \cap \mathcal{R} = \{0\}. \tag{6.89}$$
Indeed if $v \in \mathbb{H}_0^1(\Omega; \Gamma_0)$, there exists $v_n \in C_0^1(\Omega; \Gamma_0)$ such that
$$v_n \to v \quad \text{in } \mathbb{H}_0^1(\Omega; \Gamma_0).$$
Due to the trace theorem (see [17]) this implies that
$$v = 0 \quad \text{a.e. on } \Gamma_0.$$
Since $v \in \mathcal{R}$ also we obtain that
$$v = a + Ax = 0$$
on Γ_0 i.e., for every $i = 1, \ldots, n$ if $A = (a_{ij})$, $a = (a_i)$,
$$a_i + \sum_{j=1}^{n} a_{ij} x_j = 0 \quad \text{on } \Gamma_0. \tag{6.90}$$
The intersection of the hyperplanes defined by equation (6.90) contains a subset of $n - 1$-dimensional measure positive. This is only possible if these hyperplanes are all identical. But then since A is skew symmetric – i.e., the diagonal elements of A are zeros – we get that $A = 0$ and then of course $a = 0$. This shows (6.89). But then it follows from Theorem 6.11 that on $\mathbb{H}_0^1(\Omega; \Gamma_0)$ the norms
$$|u|_{1,2}, \qquad ||\varepsilon(u)||_{2,\Omega}$$

are equivalent. Next, due to Theorem 1.4, the norms

$$\|\nabla u\|_{2,\Omega}, \qquad |u|_{1,2}$$

are also equivalent on $\mathbb{H}_0^1(\Omega;\Gamma_0)$. This completes the proof of the corollary. $\square$

REMARK 6.3. The Theorems 6.9, 6.11 are also true for domains Ω with Lipschitz boundaries. See for instance [**26**], [**27**].

6.2. Existence results for linear elliptic systems

6.2.1. The case of Legendre conditions. Let Ω be a bounded Lipschitz domain of $\mathbb{R}^n$. Consider then for $i,j = 1,\ldots,n$, $\alpha,\beta = 1,\ldots,N$ coefficients $A_{\alpha\beta}^{ij} = A_{\alpha\beta}^{ij}(x) \in L^\infty(\Omega)$ satisfying for some positive constants λ, Λ

(L) $A_{\alpha\beta}^{ij}(x)\xi_i^\alpha\xi_j^\beta \geq \lambda|\xi|^2 \qquad \forall \xi \in \mathbb{R}^{n\times N}, \qquad \text{a.e. } x \in \Omega, \qquad (6.91)$

$|A_{\alpha\beta}^{ij}(x)| \leq \Lambda \qquad \forall i,j,\alpha,\beta, \qquad \text{a.e. } x \in \Omega. \qquad (6.92)$

Moreover, denote by $a_{\alpha\beta} = a_{\alpha\beta}(x)$ functions in $L^\infty(\Omega)$ such that for some non-negative constants λ', Λ' it holds that

$$a_{\alpha\beta}(x)\eta^\alpha\eta^\beta \geq \lambda'|\eta|^2 \qquad \forall \eta \in \mathbb{R}^n, \qquad \text{a.e. } x \in \Omega, \qquad (6.93)$$

$$|a_{\alpha\beta}(x)| \leq \Lambda' \qquad \forall \alpha,\beta = 1,\ldots,N, \qquad \text{a.e. } x \in \Omega. \qquad (6.94)$$

If Γ_0 is a subset of the boundary of Ω of positive measure, we denote by $\mathbb{H}_0^1(\Omega;\Gamma_0)$ the space

$$\mathbb{H}_0^1(\Omega;\Gamma_0) = H_0^1(\Omega;\Gamma_0)^N. \qquad (6.95)$$

Moreover, we denote by $f_1,\ldots,f_N$, N elements of $H_0^1(\Omega;\Gamma_0)'$ the dual of $H_0^1(\Omega;\Gamma_0)$ – i.e., we assume

$$f_1,\ldots,f_N \in H_0^1(\Omega;\Gamma_0)'. \qquad (6.96)$$

Then under the above assumptions we have

THEOREM 6.13. *There exists a unique* $u \in \mathbb{H}_0^1(\Omega;\Gamma_0)$ *solution to*

$$\int_\Omega A_{\alpha\beta}^{ij}\partial_{x_j}u^\beta\partial_{x_i}v^\alpha + a_{\alpha\beta}u^\beta v^\alpha\, dx = \langle f_\gamma, v^\gamma\rangle \quad \forall v \in \mathbb{H}_0^1(\Omega;\Gamma_0). \qquad (6.97)$$

(In the above equality we use the convention of summation of repeated indices.)

PROOF. It is enough to apply the Lax–Milgram theorem with $H = \mathbb{H}_0^1(\Omega;\Gamma_0)$,

$$a(u,v) = \int_\Omega A_{\alpha\beta}^{ij}\partial_{x_j}u^\beta\partial_{x_i}v^\alpha + a_{\alpha\beta}u^\beta v^\alpha\, dx,$$
$$\langle f, v\rangle = \langle f_\gamma, v^\gamma\rangle.$$

The continuity of a follows from the assumptions (6.92), (6.94). The coerciveness of a follows from (6.91), (6.93) and Theorem 1.4. Note that here we can assume $\lambda' = 0$. This completes the proof of the theorem. $\square$

COROLLARY 6.14. *Assume the conditions of Theorem 6.13. Moreover, let $\varphi \in \mathbb{H}^1(\Omega) = H^1(\Omega)^N$. Then there exists a unique solution to*

$$\begin{cases} u - \varphi \in \mathbb{H}^1_0(\Omega; \Gamma_0), \\ \displaystyle\int_\Omega A^{\alpha\beta}_{ij} \partial_{x_j} u^\beta \partial_{x_i} v^\alpha + a_{\alpha\beta} u^\beta v^\alpha \, dx = \langle f_\gamma, v^\gamma \rangle \quad \forall v \in \mathbb{H}^1_0(\Omega; \Gamma_0). \end{cases} \tag{6.98}$$

PROOF. This is an immediate consequence of the fact that u is a solution to (6.98) iff $w = u - \varphi$ is a solution to

$$\int_\Omega A^{\alpha\beta}_{ij} \partial_{x_j} w^\beta \partial_{x_i} v^\alpha + a_{\alpha\beta} w^\beta v^\alpha \, dx = \langle f_\gamma, v^\gamma \rangle - \int_\Omega A^{\alpha\beta}_{ij} \partial_{x_j} \varphi^\beta \partial_{x_i} v^\alpha + a_{\alpha\beta} \varphi^\beta v^\alpha \, dx$$

and the fact that the right-hand side of the above equality is a linear continuous form on $\mathbb{H}^1_0(\Omega; \Gamma_0)$. This completes the proof of the corollary. $\qquad\square$

REMARK 6.4. u solution to (6.98) is a solution to a mixed problem. We leave to the reader the task to write down the boundary conditions (see Chapter 1). Note also that we can choose different Γ_0 for the different components of u – i.e., we can replace $\mathbb{H}^1_0(\Omega; \Gamma_0)$ by

$$H^1_0(\Omega; \Gamma_1) \times \cdots \times H^1_0(\Omega; \Gamma_N)$$

where the Γ_i are different subsets of the boundary of Γ of positive measure.

In the case of $\mathbb{H}^1(\Omega) = H^1(\Omega)^N$ we have

THEOREM 6.15. *Assume the conditions of Theorem 6.13 with*

$$f_1, \ldots, f_N \in H^1(\Omega)' \tag{6.99}$$

and $\lambda' > 0$. Then, there exists a unique u solution to

$$\begin{cases} u \in \mathbb{H}^1(\Omega), \\ \displaystyle\int_\Omega A^{ij}_{\alpha\beta} \partial_{x_j} u^\beta \partial_{x_i} v^\alpha + a_{\alpha\beta} u^\beta v^\alpha \, dx = \langle f_\gamma, v^\gamma \rangle \quad \forall v \in \mathbb{H}^1(\Omega). \end{cases} \tag{6.100}$$

PROOF. The coerciveness of a follows here from the fact that $\lambda' > 0$. This completes the proof of the theorem. $\qquad\square$

6.2.2. The case of the Legendre–Hadamard condition.

We adopt here the notation of Section 6.2.1 – i.e., we consider coefficients $A^{ij}_{\alpha\beta}$, $a_{\alpha\beta}$ satisfying (6.92)–(6.94) but we replace (6.91) by the Legendre–Hadamard condition – i.e., we suppose that

(L–H) $\quad A^{ij}_{\alpha\beta}(x) \xi_i \xi_j \eta^\alpha \eta^\beta \geq \lambda |\xi|^2 |\eta|^2 \quad \forall \xi \in \mathbb{R}^n, \ \forall \eta \in \mathbb{R}^N, \text{ a.e. } x \in \Omega. \quad (6.101)$

We denote then by V one of the spaces

$$V = H^1_0(\Omega), \quad H^1_0(\Omega; \Gamma_0), \quad H^1(\Omega) \tag{6.102}$$

where Γ_0 is a part of positive measure of the boundary of Ω and we set

$$\mathbb{V} = V^N. \tag{6.103}$$

Then we have

THEOREM 6.16. *Let us assume* (6.92)–(6.94), (6.101). *Moreover, let us consider*

$$f_1, \ldots, f_N \in V'. \tag{6.104}$$

Then, if for every i, j, α, β

$$A^{ij}_{\alpha\beta}(x) \text{ is continuous on } \overline{\Omega}, \ \lambda' \text{ large enough,} \tag{6.105}$$

or

$$A^{ij}_{\alpha\beta} \text{ is constant on } \Omega \text{ and } V = H^1_0(\Omega), \tag{6.106}$$

then there exists a unique u solution to

$$\begin{cases} u \in \mathbb{V}, \\ \displaystyle\int_\Omega A^{ij}_{\alpha\beta} \partial_{x_j} u^\beta \partial_{x_i} v^\alpha + a_{\alpha\beta} u^\beta v^\alpha \, dx = \langle f_\gamma, v^\gamma \rangle \quad \forall\, v \in \mathbb{V}. \end{cases} \tag{6.107}$$

PROOF. This is an immediate consequence of Proposition 6.2 or the Gårding inequality. $\qquad\square$

6.2.3. The case of systems of elasticity type. We would like to consider in this section elliptic systems for which the ellipticity condition (L) holds only for symmetric matrices. As we will see, this kind of system is the archetype of the system of elasticity. Since in (L), (ξ^α_i) is supposed to be symmetric we assume here that

$$n = N. \tag{6.108}$$

Then, we denote by $A^{ij}_{\alpha\beta}(x)$, $a_{\alpha\beta}(x)$ functions in $L^\infty(\Omega)$ satisfying (6.92)–(6.94) and also

$$A^{ij}_{\alpha\beta}(x) = A^{ji}_{\beta\alpha}(x) = A^{\alpha j}_{i\beta}(x) \quad \forall\, i, j, \alpha, \beta = 1, \ldots, n, \text{ a.e. } x \in \Omega, \tag{6.109}$$

and we replace (L) by

$$A^{ij}_{\alpha\beta}(x)\xi^\alpha_i \xi^\beta_j \geq \lambda |\xi|^2 \quad \forall\, \xi \in M_n, \quad {}^t\xi = \xi, \quad \text{a.e. } x \in \Omega. \tag{6.110}$$

(M_n denotes the space of $n \times n$ matrices $\xi = (\xi^j_i)$ – where i is the index of lines and j the index of columns.)

Assuming that Ω is Lipschitz continuous and that Γ_0 denotes a subset of positive measure on $\partial\Omega$ we set

$$V = H^1_0(\Omega), \quad H^1_0(\Omega; \Gamma_0) \quad \text{or} \quad H^1(\Omega) \tag{6.111}$$

and as in the previous paragraph denote by $\mathbb{V}$ the space

$$\mathbb{V} = V^n. \tag{6.112}$$

Then we have

THEOREM 6.17. *Under the above assumptions and for*

$$f_1, \ldots, f_n \in V', \tag{6.113}$$

$$\lambda' > 0 \quad \text{if} \quad V = H^1(\Omega) \tag{6.114}$$

there exists a unique u solution to

$$\begin{cases} u \in \mathbb{V}, \\ \displaystyle\int_\Omega A_{\alpha\beta}^{ij}\partial_{x_j}u^\beta\partial_{x_i}v^\alpha + a_{\alpha\beta}u^\beta v^\alpha \, dx = \langle f_\gamma, v^\gamma\rangle \quad \forall v \in \mathbb{V}. \end{cases} \tag{6.115}$$

PROOF. It is enough to show the coerciveness on $\mathbb{V}$ of the bilinear form

$$a(u,v) = \int_\Omega A_{\alpha\beta}^{ij}\partial_{x_j}u^\beta\partial_{x_i}v^\alpha + a_{\alpha\beta}u^\beta v^\alpha \, dx. \tag{6.116}$$

The fundamental remark is that due to (6.109) we have

$$\begin{aligned} A_{\alpha\beta}^{ij}&\left(\frac{\xi_i^\alpha + \xi_\alpha^i}{2}\right)\left(\frac{\zeta_j^\beta + \zeta_\beta^j}{2}\right) \\ &= \frac{1}{4}A_{\alpha\beta}^{ij}\xi_i^\alpha\zeta_j^\beta + \frac{1}{4}A_{\alpha\beta}^{ij}\xi_\alpha^i\zeta_j^\beta + \frac{1}{4}A_{\alpha\beta}^{ij}\xi_i^\alpha\zeta_\beta^j + \frac{1}{4}A_{\alpha\beta}^{ij}\xi_\alpha^i\zeta_\beta^j \\ &= A_{\alpha\beta}^{ij}\xi_i^\alpha\zeta_j^\beta \quad \forall \xi, \zeta \in M_n. \end{aligned} \tag{6.117}$$

(Recall our summation convention.)

Applying this with $\zeta = \xi = \nabla u$ and recalling that

$$\varepsilon_{ij}(u) = \varepsilon_i^j(u) = \frac{1}{2}\{\partial_{x_j}u^i + \partial_{x_i}u^j\}$$

we obtain easily

$$a(u,u) = \int_\Omega A_{\alpha\beta}^{ij}\varepsilon_j^\beta(u)\varepsilon_i^\alpha(u) + a_{\alpha\beta}u^\beta u^\alpha \, dx \geq \lambda||\varepsilon(u)||_{2,\Omega}^2 + \lambda'||u||_{2,\Omega}. \tag{6.118}$$

If $\lambda' > 0$, the coerciveness of a follows then from the Korn inequality (6.61). When $\lambda' = 0$, it follows from Corollary 6.12. This completes the proof of the theorem. $\square$

Let us remark that in the case where $\lambda' = 0$ but

$$a_{\alpha\beta}\xi^\alpha\xi^\beta \neq 0, \tag{6.119}$$

we can develop coerciveness results in the spirit of Theorem 1.8. We will not pursue this direction further but consider now the case where

$$a_{\alpha\beta} \equiv 0. \tag{6.120}$$

Then let us denote by $\mathbb{V}$ a closed subspace of $\mathbb{H}^1(\Omega) = H^1(\Omega)^n$. Let

$$f_1, f_2, \ldots, f_N \in H^1(\Omega)'. \tag{6.121}$$

Then we have:

THEOREM 6.18. *Assuming* (6.92), (6.94), (6.109), (6.110), (6.121) *if*

$$\mathbb{V} \cap \mathcal{R} = \{0\}, \tag{6.122}$$

then there exists a unique u solution to

$$\begin{cases} u \in \mathbb{V}, \\ \displaystyle\int_\Omega A_{\alpha\beta}^{ij}\partial_{x_j}u^\beta\partial_{x_i}v^\alpha \, dx = \langle f, v\rangle = \langle f_\gamma, v^\gamma\rangle \quad \forall v \in \mathbb{V}. \end{cases} \tag{6.123}$$

If

$$\mathbb{V} \cap \mathcal{R} = \mathcal{R}_1 \neq \{0\}, \tag{6.124}$$

then there exists a solution to (6.123) iff it holds that

$$\langle f, r_1 \rangle = 0 \quad \forall r_1 \in \mathcal{R}_1. \tag{6.125}$$

Moreover if u is a solution to (6.123) every other solution is of the type $u + r_1$, $r_1 \in \mathcal{R}_1$.

PROOF. If $\mathbb{V} \cap \mathcal{R} = \{0\}$, then by Theorem 6.11 and (6.117) it holds that

$$
\begin{aligned}
a(u, u) &= \int_\Omega A_{\alpha\beta}^{ij} \partial_{x_j} u^\beta \partial_{x_i} u^\alpha \, dx \\
&= \int_\Omega A_{\alpha\beta}^{ij} \varepsilon_j^\beta(u) \varepsilon_i^\alpha(u) \, dx \\
&\geq \lambda \|\varepsilon(u)\|_{2,\Omega}^2 \geq C|u|_{1,2}^2 \quad \forall u \in \mathbb{V},
\end{aligned}
\tag{6.126}
$$

for some constant C and since a is coercive on $\mathbb{V}$ the result follows from the Lax–Milgram theorem.

Suppose now that (6.124) holds. Then if u is a solution to (6.123), by (6.117) it holds that

$$\int_\Omega A_{\alpha\beta}^{ij} \partial_{x_j} u^\beta \partial_{x_i} r_1^\alpha \, dx = \int_\Omega A_{\alpha\beta}^{ij} \varepsilon_j^\beta(u) \varepsilon_j^\beta(r_1) \, dx = 0 = \langle f, r_1 \rangle \quad \forall r_1 \in \mathcal{R}_1. \tag{6.127}$$

Thus (6.125) holds. Conversely, let us assume that (6.124), (6.125) hold. Denote by $\mathbb{V}_1$ the orthogonal of $\mathcal{R}_1$ in $\mathbb{V}$ for the usual scalar product in $\mathbb{H}^1(\Omega)$. If $v \in \mathbb{V}_1 \cap \mathcal{R}$, then $v \in \mathbb{V}_1$ and $v \in \mathcal{R}_1$ thus $v = 0$ and we have

$$\mathbb{V}_1 \cap \mathcal{R} = \{0\}.$$

From the first part of the theorem it follows that there exists $u \in \mathbb{V}_1$ unique such that

$$\int_\Omega A_{\alpha\beta}^{ij} \partial_{x_j} u^\beta \partial_{x_i} v^\alpha \, dx = \langle f, v \rangle \quad \forall v \in \mathbb{V}_1.$$

Due to (6.127) we have also

$$\int_\Omega A_{\alpha\beta}^{ij} \partial_{x_j} u^\beta \partial_{x_i} v^\alpha \, dx = \langle f, v \rangle \quad \forall v \in \mathbb{V}_1 + \mathcal{R}_1 = \mathbb{V}$$

and a solution to (6.123) exists. Clearly $u + r_1$ is also a solution for every $r_1 \in \mathcal{R}_1$ and we have obtained in this way all the solutions to (6.123). This completes the proof of the theorem. $\square$

As a corollary we have

COROLLARY 6.19. *Under the assumptions of Theorem 6.18 the problem*

$$\begin{cases} u \in \mathbb{H}^1(\Omega), \\ \displaystyle\int_\Omega A_{\alpha\beta}^{ij} \partial_{x_j} u^\alpha \partial_{x_i} v^\alpha \, dx = \langle f, v \rangle = \langle f_\gamma, v^\gamma \rangle \quad \forall v \in \mathbb{H}^1(\Omega) \end{cases} \tag{6.128}$$

has a solution for $f_1, \ldots, f_n \in H^1(\Omega)'$ if and only if

$$\langle f, r \rangle = 0 \quad \forall r \in \mathcal{R}. \tag{6.129}$$

Moreover if u is a solution to (6.128) then any other solution is given by $u + r$ with $r \in \mathcal{R}$.

PROOF. It is enough to apply Theorem 6.18 with $\mathbb{V} = \mathbb{H}^1(\Omega)$. Then $\mathcal{R}_1 = \mathcal{R}$.
□

REMARK 6.5. Due to the symmetry condition (6.109), (see also (6.117)) we have

$$A^{ij}_{\alpha\beta}\xi^\beta_j\xi^\alpha_i = A^{ij}_{\alpha\beta}\left(\frac{\xi^\beta_j + \xi^j_\beta}{2}\right)\left(\frac{\xi^\alpha_i + \xi^i_\alpha}{2}\right) \geq \lambda\left|\frac{\xi + {}^t\xi}{2}\right|^2 \quad \forall \xi \in M_n, \tag{6.130}$$

since the quadratic form that we are considering is coercive on the space of symmetric matrices. However, it could vanish for a skew symmetric matrix so that the Legendre condition does not hold. If in (6.130) we consider

$$\xi^\beta_j = \varphi_j \eta^\beta$$

we obtain

$$
\begin{aligned}
A^{ij}_{\alpha\beta}\varphi_j\eta^\beta\varphi_i\eta^\alpha &\geq \frac{\lambda}{4}\sum_{j,\beta}(\varphi_j\eta^\beta + \varphi_\beta\eta^j)^2 \\
&= \frac{\lambda}{4}\sum_{j,\beta}\varphi_j^2(\eta^\beta)^2 + \varphi_\beta^2(\eta^j)^2 + 2\varphi_j\eta^j\varphi_\beta\eta^\beta \\
&= \frac{\lambda}{4}\{|\varphi|^2|\eta|^2 + |\varphi|^2|\eta|^2 + 2(\varphi \cdot \eta)^2\} \\
&\geq \frac{\lambda}{2}|\varphi|^2|\eta|^2
\end{aligned}
\tag{6.131}
$$

with an obvious notation for $|\varphi|, |\eta|$. Thus the Legendre–Hadamard condition holds in this case.

The most famous application of systems satisfying (6.109) is the system of linear elasticity. For this system we have (see [25], [15], [16])

$$A^{ij}_{\alpha\beta} = \lambda\delta^i_\alpha\delta^j_\beta + \mu\delta^i_\beta\delta^j_\alpha + \mu\delta^i_j\delta^\alpha_\beta \tag{6.132}$$

where λ, μ are two positive constants (we could have them depending on x as well) called the Lamé constants and δ^i_α denotes the Kronecker symbol defined as

$$
\begin{aligned}
\delta^i_\alpha =& 1 \quad \text{if } i = \alpha, \\
& 0 \quad \text{if } i \neq \alpha.
\end{aligned}
\tag{6.133}
$$

It is clear that the symmetry relation (6.109) holds for $A^{ij}_{\alpha\beta}$ given by (6.132). Moreover if $\xi = (\xi^\beta_j)$, $\varphi = (\varphi^\alpha_i) \in M_n$, tr denotes the trace of matrices, it holds that

$$A^{ij}_{\alpha\beta}\xi^{\beta}_{j}\varphi^{\alpha}_{i} = \lambda\xi^{j}_{j}\varphi^{i}_{i} + \mu\xi^{i}_{j}\varphi^{j}_{i} + \mu\xi^{j}_{i}\varphi^{j}_{i}$$
$$= \lambda\operatorname{tr}\xi\operatorname{tr}\varphi + \mu(\xi + {}^{t}\xi):\varphi \tag{6.134}$$

where $\xi : \varphi$ denotes the usual scalar product of matrices – i.e., the euclidean scalar product for the vectors formed by the coefficients. Since $\xi + {}^{t}\xi$ is symmetric we have also

$$(\xi + {}^{t}\xi) : \varphi = (\xi + {}^{t}\xi) : {}^{t}\varphi$$

and (6.134) has also the more symmetric form

$$A^{ij}_{\alpha\beta}\xi^{\beta}_{j}\varphi^{\alpha}_{i} = \lambda\operatorname{tr}\xi\operatorname{tr}\varphi + 2\mu\left(\frac{\xi + {}^{t}\xi}{2}\right):\left(\frac{\varphi + {}^{t}\varphi}{2}\right). \tag{6.135}$$

Let us now translate the existence and uniqueness results that we have obtained in the framework of elasticity. We assume in the rest of this section that $n = 3$ only to restrict ourselves to the physical situation and consider a body occupying a domain Ω. This body is supposed to be maintained or clamped on a part Γ_0 of its

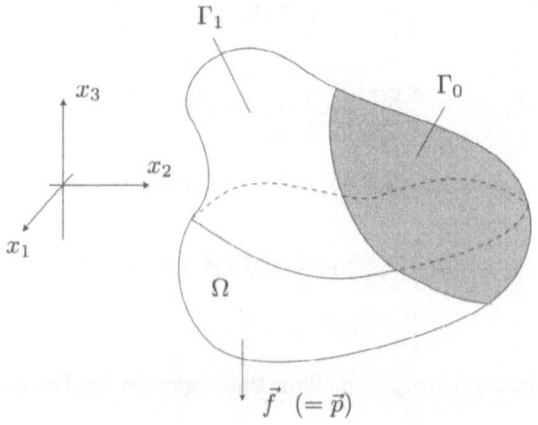

FIGURE 6.3

boundary. Some density of forces (for instance its weight) is acting inside Ω and on the part $\Gamma_1 = \partial\Omega \setminus \Gamma_0$. The problem is then to determine the deformation of this body subject to these forces when a stationary deformation has been achieved. We denote by

$$f = (f_1, f_2, f_3) \quad \text{the } density \text{ of forces inside } \Omega, \tag{6.136}$$

$$g = (g_1, g_2, g_3) \quad \text{the } density \text{ of forces acting on } \Gamma_1, \tag{6.137}$$

and to simplify our exposition we suppose that

$$f \in \mathbb{L}^2(\Omega), \qquad g \in \mathbb{L}^2(\Gamma_1) \tag{6.138}$$

with an obvious notation for $\mathbb{L}^2(\Omega)$, $\mathbb{L}^2(\Gamma_1)$. Then we set

$$\langle f, v \rangle = \int_\Omega f_i v^i \, dx + \int_{\Gamma_1} g_i v^i d\gamma(x) \quad \forall v \in \mathbb{H}^1(\Omega). \tag{6.139}$$

It is clear that $\langle f, \cdot \rangle$ defined this way is a continuous linear form on $\mathbb{H}^1(\Omega) = (H^1(\Omega))^3$. ($d\gamma$ denotes the measure area on $\partial\Omega$ the boundary of Ω.) Thus we have

THEOREM 6.20. *Under the above assumption, for* $\lambda \geq 0$, $\mu > 0$, *and assuming* $|\Gamma_0| \neq 0$, *there exists a unique u solution to*

$$\begin{cases} u \in \mathbb{H}_0^1(\Omega; \Gamma_0) = H_0^1(\Omega; \Gamma_0)^3, \\ \lambda \int_\Omega \operatorname{div} u \operatorname{div} v \, dx + 2\mu \int_\Omega \varepsilon_{ij}(u)\varepsilon_{ij}(v) \, dx = \langle f, v \rangle \\ \forall v \in \mathbb{H}_0^1(\Omega; \Gamma_0). \end{cases} \tag{6.140}$$

Moreover setting

$$\sigma_{ij}(u) = \lambda \operatorname{div} u \delta_i^j + 2\mu\varepsilon_{ij}(u) \tag{6.141}$$

u is a weak solution to

$$-\partial_{x_j}\sigma_{ij}(u) = f_i \quad i = 1, 2, 3 \quad in \quad \Omega, \tag{6.142}$$

$$u = 0 \quad on \ \Gamma_0, \qquad \sigma_{ij}(u)\nu_j = g_i \quad on \quad \Gamma_1. \tag{6.143}$$

PROOF. Due to (6.135) we have for any symmetric matrix ξ

$$A_{\alpha\beta}^{ij}\xi_j^\beta\xi_i^\alpha \geq \lambda(\operatorname{tr}\xi)^2 + 2\mu|\xi|^2 \geq 2\mu|\xi|^2. \tag{6.144}$$

Thus we can apply Theorem 6.17 to get existence and uniqueness of a solution u to

$$\begin{cases} u \in \mathbb{H}_0^1(\Omega; \Gamma_0), \\ \int_\Omega A_{\alpha\beta}^{ij}\partial_{x_j}u^\beta\partial_{x_i}v^\alpha \, dx = \int_\Omega f_\alpha v^\alpha \, dx + \int_{\Gamma_1} g_\alpha v^\alpha \, d\gamma(x) = \langle f, v \rangle \\ \forall v \in \mathbb{H}_0^1(\Omega; \Gamma_0). \end{cases} \tag{6.145}$$

Taking into account (6.135) u is a solution to (6.140). Moreover, by (6.134) u is also a solution to

$$\int_\Omega \{\lambda \operatorname{div} u \operatorname{div} v + 2\mu\varepsilon_{ij}(u)\partial_{x_j}v^i\} \, dx$$
$$= \int_\Omega f_\alpha v^\alpha \, dx + \int_{\Gamma_1} g_\alpha v^\alpha \, d\gamma(x) \quad \forall v \in H_0^1(\Omega; \Gamma_0). \tag{6.146}$$

Taking $v \in \mathcal{D}(\Omega)^3$ in the above equation we derive that

$$-\partial_{x_j}\{\lambda \operatorname{div} u \delta_i^j + 2\mu\varepsilon_{ij}(u)\} = f_i \quad i = 1, 2, 3 \text{ in } \Omega,$$

$$\Rightarrow \qquad -\partial_{x_j}\{\sigma_{ij}(u)\} = f_i \quad i = 1, 2, 3 \text{ in } \Omega. \tag{6.147}$$

Writing down (6.146) under the form

$$\int_\Omega \sigma_{ij}(u)\partial_{x_j}v^i \, dx = \int_\Omega f_\alpha v^\alpha \, dx + \int_{\Gamma_1} g_\alpha v^\alpha \, d\gamma(x)$$

and assuming the functions above to be smooth, we get after integrating by parts

$$\int_\Omega \left[\partial_{x_j} \{ \sigma_{ij}(u)v^i \} - \partial_{x_j}(\sigma_{ij}(u))v^i - f_i v^i \right] dx = \int_{\Gamma_1} g_\alpha v^\alpha \, d\gamma(x)$$

for every smooth function $v \in \mathbb{H}_0^1(\Omega; \Gamma_0)$. From (6.147) we derive

$$\int_{\Gamma_1} \sigma_{ij}(u)\nu_j v^i \, d\gamma(x) = \int_{\Gamma_1} g_i v^i \, d\gamma(x)$$

which leads to the weak formulation (6.142), (6.143). This completes the proof of the theorem. □

Thus we have found the displacement u of our body subject to the density of forces f and g (for a physical background see for instance [**25**]). To end this section we address the so-called *"pure traction problem"* i.e., the case where

$$|\Gamma_0| = 0.$$

Then, as an immediate consequence of Corollary 6.19 we have

THEOREM 6.21. *Suppose that we are under the assumptions of Theorem 6.20. Then if $\langle f, \cdot \rangle$ is defined by (6.139) and*

$$\langle f, r \rangle = 0 \quad \forall r \in \mathcal{R} \tag{6.148}$$

there exists u solution to

$$\begin{cases} u \in \mathbb{H}^1(\Omega), \\ \lambda \int_\Omega \operatorname{div} u \operatorname{div} v \, dx + 2\mu \int_\Omega \varepsilon_{ij}(u)\varepsilon_{ij}(v) \, dx = \langle f, v \rangle \quad \forall v \in \mathbb{H}^1(\Omega). \end{cases} \tag{6.149}$$

Moreover, u is uniquely determined up to a rigid motion and satisfies in a weak sense

$$\begin{cases} -\partial_{x_j} \{ \sigma_{ij}(u) \} = f_i & in \ \Omega, \ i = 1, 2, 3, \\ \sigma_{ij}(u)\nu_j = g_i & on \ \Gamma = \partial\Omega, \ i = 1, 2, 3. \end{cases} \tag{6.150}$$

PROOF. The proof follows the lines of Theorem 6.20 and is left to the reader. It is clear that u solution to (6.149) or (6.150) is the displacement inside a body occupying the *reference configuration* Ω and subjected to forces f, g. This displacement is of course determined up to a rigid motion. □

REMARK 6.6. Theorem 6.20 and 6.21 hold true with the same arguments for any n. In particular for $n = 2$ it gives us results for two dimensional elasticity. The details are left to the reader.

6.3. Nonlinear elliptic systems

We conclude this chapter with a few nonlinear results. Let Ω denote a bounded open subset of $\mathbb{R}^n$ with a Lipschitz boundary. Let us consider for $i, j = 1, \ldots, n$, $\alpha, \beta = 1, \ldots, N$, Carathéodory functions $A_{\alpha\beta}^{ij}(x, u)$. We recall that this means that

$$x \mapsto A_{\alpha\beta}^{ij}(x, u) \quad \text{is measurable } \forall u \in \mathbb{R}^N, \tag{6.151}$$

$$u \mapsto A_{\alpha\beta}^{ij}(x, u) \quad \text{is continuous a.e. } x \in \Omega. \tag{6.152}$$

Moreover let us assume that for some positive constants λ, Λ it holds that

$$|A_{\alpha\beta}^{ij}(x, u)| \leq \Lambda \quad \text{a.e. } x \in \Omega, \ \forall u \in \mathbb{R}^N,$$
$$\forall i, j = 1, \ldots, n, \ \forall \alpha, \beta = 1, \ldots, N, \tag{6.153}$$

$$A_{\alpha\beta}^{ij}(x, u)\xi_i^\alpha \xi_j^\beta \geq \lambda |\xi|^2 \quad \forall \xi \in \mathbb{R}^{n \times N}, \text{ a.e. } x \in \Omega, \ \forall u \in \mathbb{R}^N. \tag{6.154}$$

If $\mathbb{H}_0^1(\Omega; \Gamma_0)$ denote the space defined by (6.95), (with $|\Gamma_0| > 0$), and if $f_1, \ldots, f_N$ are N elements of its dual space – i.e., if

$$f_1, f_2, \ldots, f_N \in H_0^1(\Omega; \Gamma_0)' \tag{6.155}$$

then we have

THEOREM 6.22. *Under the above assumptions there exists* $u \in \mathbb{H}_0^1(\Omega; \Gamma_0)$ *solution to*

$$\int_\Omega A_{\alpha\beta}^{ij}(x, u(x)) \partial_{x_j} u^\beta \partial_{x_i} v^\alpha \, dx = \langle f_\gamma, v^\gamma \rangle \quad \forall v \in \mathbb{H}_0^1(\Omega; \Gamma_0). \tag{6.156}$$

PROOF. It is enough to use the Schauder fixed point theorem as we already did in Theorem 4.4. More precisely it results from Theorem 6.13 and (6.151)–(6.154) that for every $w \in \mathbb{L}^2(\Omega) = L^2(\Omega)^N$ there exists a unique u solution to

$$\begin{cases} u \in \mathbb{H}_0^1(\Omega; \Gamma_0), \\ \int_\Omega A_{\alpha\beta}^{ij}(x, w(x)) \partial_{x_j} u^\beta \partial_{x_i} v^\alpha \, dx = \langle f_\gamma, v^\gamma \rangle \quad \forall v \in \mathbb{H}_0^1(\Omega; \Gamma_0). \end{cases} \tag{6.157}$$

If we can show that this mapping $w \to u = T(w)$ has a fixed point we are done. For that, taking $v = u$ in the second equation (6.157) we get from (6.154) for some constant

$$||\nabla u||_{2,\Omega}^2 \leq C \tag{6.158}$$

where C is independent of w. It follows then from the Poincaré inequality that for some other constant C it holds also that

$$||u||_{2,\Omega}^2 \leq C. \tag{6.159}$$

Thus, T is a mapping from $B(0, C)$, the ball of center 0 and radius C in $\mathbb{L}^2(\Omega)$, into itself. Moreover, by (6.158), $T(B)$ is relatively compact in B. One can then establish as in Theorem 4.4 the continuity of T. This completes the proof by the Schauder fixed point theorem. $\square$

REMARK 6.7. Of course we could include lower order terms in the operator. For the sake of simplicity we avoid this refinement.

A nonlinear theory of elasticity is also available. To see this consider for $n = N$ Carathéodory functions $A_{\alpha\beta}^{ij}(x, u)$ satisfying (6.151)–(6.153) with in addition the symmetry condition

$$A_{\alpha\beta}^{ij}(x, u) = A_{\beta\alpha}^{ji}(x, u) = A_{i\beta}^{\alpha j}(x, u) \quad \text{a.e. } x \in \Omega, \ \forall u \in \mathbb{R}^n. \tag{6.160}$$

Moreover, let us replace the Legendre condition (6.154) by

$$A_{\alpha\beta}^{ij}(x, u)\xi_i^\alpha \xi_j^\beta \geq \lambda |\xi|^2 \quad \forall \xi \in M_n, \ {}^t\xi = \xi, \ \text{a.e. } x \in \Omega. \tag{6.161}$$

Then we have

THEOREM 6.23. *Let $f_1, f_2, \ldots, f_n$ be such that*

$$f_1, f_2, \ldots, f_n \in \mathbb{H}_0^1(\Omega; \Gamma_0)'. \tag{6.162}$$

Under the above assumptions there exists a solution u to

$$\begin{cases} u \in \mathbb{H}_0^1(\Omega; \Gamma_0), \\ \displaystyle\int_\Omega A_{\alpha\beta}^{ij}(x, u(x)) \partial_{x_j} u^\beta \partial_{x_i} v^\alpha \, dx = \langle f_\gamma, v^\gamma \rangle \quad \forall v \in \mathbb{H}_0^1(\Omega; \Gamma_0). \end{cases} \tag{6.163}$$

PROOF. We proceed as in Theorem 6.22 introducing for $w \in \mathbb{L}^2(\Omega)$, $u = T(w)$ solution to

$$\begin{cases} u \in \mathbb{H}_0^1(\Omega; \Gamma_0), \\ \displaystyle\int_\Omega A_{\alpha\beta}^{ij}(x, w(x)) \partial_{x_j} u^\beta \partial_{x_i} v^\alpha \, dx = \langle f_\gamma, v^\gamma \rangle \quad \forall v \in \mathbb{H}_0^1(\Omega; \Gamma_0). \end{cases} \tag{6.164}$$

Taking $v = u$ in the above equation we derive by (6.161) that

$$\|\varepsilon(u)\|_{2,\Omega}^2 \leq C$$

where C is independent of w. It follows then from Corollary 6.12 that

$$\|\nabla u\|_{2,\Omega}^2 \leq C.$$

One can then conclude exactly as in Theorem 6.22. $\qquad\square$

Open problems

1. We do not know if the best constant C in (6.32) is known.
2. Except in the case of diagonal systems very little is known about the uniqueness of a solution to (6.156) or (6.163).

Chapter 7

Asymptotic Behaviour of Elliptic Systems

From the point of view of applications, one of the most important cases of an elliptic system is the system of elasticity. Before attacking it we will first consider the case of linear elliptic systems satisfying the Legendre condition. For such a system we will see that the techniques developed in Chapters 2 and 3 apply with almost no change.

7.1. The case of linear elliptic systems satisfying the Legendre condition

For the sake of simplicity we will restrict ourselves to the case of Dirichlet boundary conditions. However, the reader will have no trouble showing that our method extends with almost no change to the case of mixed boundary conditions.

So, we denote for $\ell > 0$ by Ω_ℓ the open cylinder of $\mathbb{R}^n$

$$\Omega_\ell = (-\ell, \ell)^p \times \omega \tag{7.1}$$

where ω is a bounded open subset of $\mathbb{R}^{n-p}$, $1 \leq p < n$. As we are now used to, points in $\mathbb{R}^n$ are denoted by $x = (X_1, X_2)$ where

$$X_1 = (x_1, \ldots, x_p), \qquad X_2 = (x_{p+1}, \ldots, x_n). \tag{7.2}$$

For $\alpha = 1, \ldots, N$, we denote by f_α elements of $H^{-1}(\omega)$ the dual of $H_0^1(\omega)$. Then we would like to consider systems defined through coefficients $A_{\alpha\beta}^{ij}$ such that

$$A_{\alpha\beta}^{ij} = A_{\alpha\beta}^{ij}(x) \in L^\infty(\mathbb{R}^p \times \omega), \quad |A_{\alpha\beta}^{ij}(x)| \leq \Lambda \text{ a.e. } x \in \mathbb{R}^p \times \omega, \; \forall i, j, \alpha, \beta, \tag{7.3}$$

and satisfying also the Legendre condition – i.e., such that (see (6.3)) – for some positive constant λ, it holds that

$$A_{\alpha\beta}^{ij}(x)\xi_j^\beta \xi_i^\alpha \geq \lambda |\xi|^2 \quad \forall \xi = (\xi_i^\alpha) \in \mathbb{R}^{nN}, \text{ a.e. } x \in \mathbb{R}^p \times \omega. \tag{7.4}$$

(Recall that in the above formula the summation is performed in i, j, α, β as in general in what follows – for any pair of indices identical sitting at different levels.) For any $\alpha = 1, \ldots, N$, it is clear that we can define

$$\langle f_\alpha, v^\alpha \rangle \tag{7.5}$$

for every $v^\alpha \in H_0^1(\Omega_\ell)$ via the formula (3.21). Then it follows from Theorem 6.13 that there exists a unique $u_\ell = (u_\ell^1, \ldots, u_\ell^N)$ solution to

$$\begin{cases} u_\ell \in \mathbb{H}_0^1(\Omega_\ell), \\ \int_{\Omega_\ell} A_{\alpha\beta}^{ij}(x) \partial_{x_j} u_\ell^\beta \partial_{x_i} v^\alpha \, dx = \langle f_\alpha, v^\alpha \rangle \quad \forall v = (v^1, \ldots, v^N) \in \mathbb{H}_0^1(\Omega_\ell). \end{cases} \quad (7.6)$$

Recall that $\mathbb{H}_0^1(\Omega_\ell) = (H_0^1(\Omega_\ell))^N$.

We would like then to study the asymptotic behaviour of u_ℓ when $\ell \to +\infty$. For that we will make an assumption that includes our assumption made in Chapter 3 as a particular case (when $N = 1$), that is to say we will assume that

$$A_{\alpha\beta}^{ij}(x) = A_{\alpha\beta}^{ij}(X_2) \quad \forall j = p+1, \ldots, n \quad \forall i, \alpha, \beta. \quad (7.7)$$

Then, under the above assumptions, it is clear that if in (7.4) we choose (ξ_i^α) such that

$$\xi_i^\alpha = 0 \quad \forall i = 1, \ldots, p, \quad \forall \alpha = 1, \ldots, N,$$

then (7.4) becomes

$$\sum_{i,j=p+1}^n A_{\alpha\beta}^{ij}(X_2) \xi_j^\beta \xi_i^\alpha \geq \lambda |\xi|^2 \quad \forall \xi = (\xi_i^\alpha) \in \mathbb{R}^{(n-p)N}, \text{ a.e. } X_2 \in \omega. \quad (7.8)$$

(In the above formula we did not indicate the summation in α, β. Recall also that $|\xi|$ denotes the euclidean norm of vectors or matrices – i.e., in this case $|\xi|^2 = \sum_{i=p+1}^n \sum_{\alpha=1}^N (\xi_i^\alpha)^2$.) Taking into account (7.8) it follows from Theorem 6.13 that there exists a unique $u_\infty = (u_\infty^1, \ldots, u_\infty^N)$ solution to

$$\begin{cases} u_\infty \in \mathbb{H}_0^1(\omega), \\ \sum_{i,j=p+1}^n \int_\omega A_{\alpha\beta}^{ij}(X_2) \partial_{x_j} u_\infty^\beta \partial_{x_i} v^\alpha \, dX_2 = \langle f_\alpha, v^\alpha \rangle \\ \forall v = (v^1, \ldots, v^N) \in \mathbb{H}_0^1(\omega). \end{cases} \quad (7.9)$$

(Recall that $\mathbb{H}_0^1(\omega) = (H_0^1(\omega))^N$, $\langle \cdot \rangle$ is the duality bracket between $H^{-1}(\omega)$ and $H_0^1(\omega)$ and in the integral a summation in α, β occurs.) Then, as already seen in Chapter 3 in the case $N = 1$, under the above assumptions we have

THEOREM 7.1. *Under the above assumptions let u_ℓ, u_∞ be the solutions to (7.6), (7.9) respectively. Then, for any $\ell_0 > 0$, any $r > 0$ there exists a constant C independent of ℓ such that*

$$\|\nabla(u_\ell - u_\infty)\|_{2,\Omega_{\ell_0}} \leq \frac{C}{\ell^r}. \quad (7.10)$$

($\nabla(u_\ell - u_\infty)$ denotes the Jacobian matrix of $u_\ell - u_\infty$ – i.e., the matrix with entries $\partial_{x_i}(u_\ell^\alpha - u_\infty^\alpha)$, $|\cdot|$ the euclidean norm of this matrix, $|\cdot|_{2,\Omega_{\ell_0}}$ the usual $L^2(\Omega_{\ell_0})$-norm.)

Before to go into the proof of Theorem 7.1 let us start as usual by an estimate on u_ℓ generalizing the estimate (3.29). We have

LEMMA 7.2. *There exists a constant $C = C(\lambda, \Lambda, n, p)$ independent of ℓ such that*

$$||\nabla u_\ell||_{2,\Omega_\ell} \leq C\ell^{p/2}||\nabla u_\infty||_{2,\omega}. \tag{7.11}$$

PROOF. We take as test function in (7.6) $v = u_\ell$ and as test function in (7.9) $v = u_\ell(\mathrm{X}_1, \cdot)$ and integrate in X_1. It follows easily that

$$\int_{\Omega_\ell} A_{\alpha\beta}^{ij}(x)\partial_{x_j}u_\ell^\beta \partial_{x_i}u_\ell^\alpha \, dx = \sum_{i,j=p+1}^{n} \int_{\Omega_\ell} A_{\alpha\beta}^{ij}(\mathrm{X}_2)\partial_{x_j}u_\infty^\beta \partial_{x_i}u_\ell^\alpha \, dx. \tag{7.12}$$

Using the ellipticity condition and (7.3) we get for some constant $C = C(n, \Lambda)$

$$\lambda||\nabla u_\ell||_{2,\Omega_\ell}^2 \leq C||\nabla u_\infty||_{2,\Omega_\ell}||\nabla u_\ell||_{2,\Omega_\ell}.$$

From which follows

$$||\nabla u_\ell||_{2,\Omega_\ell} \leq \frac{C}{\lambda}(2\ell)^{p/2}||\nabla u_\infty||_{2,\omega}.$$

This completes the proof.   □

PROOF OF THEOREM 7.1. Let – as in Chapter 3 – ϱ be a smooth function of $\mathbb{R}^p$ such that

$$0 \leq \varrho \leq 1, \quad \varrho = 1 \quad \text{on} \quad \left(-\frac{1}{2}, \frac{1}{2}\right)^p, \quad \varrho = 0 \quad \text{on} \quad \mathbb{R}^p \setminus (-1, 1)^p, \tag{7.13}$$

$$|\nabla_{\mathrm{X}_1}\varrho| \leq \theta \tag{7.14}$$

for some constant θ. Considering for $\ell_1 \leq \ell$

$$\varrho^2\left(\frac{\mathrm{X}_1}{\ell_1}\right)(u_\ell - u_\infty) \in \mathbb{H}_0^1(\Omega_\ell) \tag{7.15}$$

as test function in (7.6) and $\varrho^2(\frac{\mathrm{X}_1}{\ell_1})(u_\ell - u_\infty)(\mathrm{X}_1, \cdot)$ as test function in (7.9) we derive that

$$\int_{\Omega_\ell} A_{\alpha\beta}^{ij}(x)\partial_{x_j}u_\ell^\beta \partial_{x_i}\{\varrho^2(u_\ell^\alpha - u_\infty^\alpha)\} \, dx$$

$$= \sum_{i,j=p+1}^{n} \int_{\Omega_\ell} A_{\alpha\beta}^{ij}(\mathrm{X}_2)\partial_{x_j}u_\infty^\beta \partial_{x_i}\{\varrho^2(u_\ell^\alpha - u_\infty^\alpha)\} \, dx. \tag{7.16}$$

Let us investigate more closely the integral on the right-hand side of (7.16). It holds that, since u_∞ depends on X_2 only,

$$\sum_{i,j=p+1}^{n} \int_{\Omega_\ell} A_{\alpha\beta}^{ij}(\mathrm{X}_2)\partial_{x_j}u_\infty^\beta \partial_{x_i}\{\varrho^2(u_\ell^\alpha - u_\infty^\alpha)\} \, dx$$

$$= \sum_{\substack{i=p+1,\dots,n \\ j=1,\dots,n}} \int_{\Omega_\ell} A_{\alpha\beta}^{ij}(x)\partial_{x_j}u_\infty^\beta \partial_{x_i}\{\varrho^2(u_\ell^\alpha - u_\infty^\alpha)\} \, dx. \tag{7.17}$$

Moreover, it holds that, since $u_\infty = u_\infty(\mathrm{x}_2)$,

$$\sum_{\substack{i=1,\ldots,p \\ j=1,\ldots,n}} \int_{\Omega_\ell} A^{ij}_{\alpha\beta}(x)\partial_{x_j} u^\beta_\infty \partial_{x_i}\{\varrho^2(u^\alpha_\ell - u^\alpha_\infty)\}\,dx$$

$$= \sum_{\substack{i=1,\ldots,p \\ j=p+1,\ldots,n}} \int_{\Omega_\ell} A^{ij}_{\alpha\beta}(\mathrm{x}_2)\partial_{x_j} u^\beta_\infty(\mathrm{x}_2)\partial_{x_i}\{\varrho^2(u^\alpha_\ell - u^\alpha_\infty)\}\,dx \qquad (7.18)$$

$$= 0.$$

(It is enough to integrate in x_1 in the last integral.) Collecting (7.16), (7.17), (7.18) we obtain

$$\int_{\Omega_\ell} A^{ij}_{\alpha\beta}(x)\partial_{x_j}(u^\beta_\ell - u^\alpha_\infty)\partial_{x_i}\{(u^\alpha_\ell - u^\alpha_\infty)\varrho^2\}\,dx = 0. \qquad (7.19)$$

Performing the derivation in x_i it follows immediately that

$$\int_{\Omega_{\ell_1}} A^{ij}_{\alpha\beta}(x)\partial_{x_j}(u^\beta_\ell - u^\alpha_\infty)\partial_{x_i}(u^\alpha_\ell - u^\alpha_\infty)\varrho^2\,dx$$

$$= -\frac{2}{\ell_1}\sum_{i=1}^p \int_{\Omega_{\ell_1}} A^{ij}_{\alpha\beta}(x)\partial_{x_j}(u^\beta_\ell - u^\alpha_\infty)\partial_{x_i}\varrho\left(\frac{\mathrm{x}_1}{\ell_1}\right)(u^\alpha_\ell - u^\alpha_\infty)\varrho\left(\frac{\mathrm{x}_1}{\ell_1}\right)dx. \qquad (7.20)$$

(Note that by (7.13) it is enough to integrate on Ω_{ℓ_1}. In this last integral, summation is understood in j, α, β.) Due to the Legendre condition (7.4) and (7.3) we derive easily for $C = C(n, \Lambda)$

$$\lambda \int_{\Omega_{\ell_1}} |\nabla(u_\ell - u_\infty)|^2\varrho^2\,dx \leq \frac{C}{\ell_1}\int_{\Omega_{\ell_1}} |\nabla(u_\ell - u_\infty)||u_\ell - u_\infty|\varrho\,dx. \qquad (7.21)$$

(Recall that $|\cdot|$ denotes the euclidean norm of vectors or matrices, ∇v is the Jacobian matrix of the vector v.) Using the Cauchy–Schwarz inequality in the last integral it follows that

$$\int_{\Omega_{\ell_1}} |\nabla(u_\ell - u_\infty)|^2\varrho^2\,dx \leq \frac{C}{\lambda\ell_1}||\nabla(u_\ell - u_\infty)||_{2,\Omega_{\ell_1}}||u_\ell - u_\infty|\varrho|_{2,\Omega_{\ell_1}}. \qquad (7.22)$$

Next we apply Lemma 3.3 for $a = 0$. We then obtain for some constant C independent of ℓ_1

$$\big|\big|\nabla(u_\ell - u_\infty)|\varrho\big|\big|^2_{2,\Omega_{\ell_1}} \leq \frac{C}{\ell_1}||\nabla(u_\ell - u_\infty)||_{2,\Omega_{\ell_1}}|\nabla(u_\ell - u_\infty)|\varrho|_{2,\Omega_{\ell_1}}$$

from which follows

$$\big|\big|\nabla(u_\ell - u_\infty)|\varrho\big|_{2,\Omega_{\ell_1}} \leq \frac{C}{\ell_1}||\nabla(u_\ell - u_\infty)||_{2,\Omega_{\ell_1}}.$$

Recalling (7.13) we have

$$||\nabla(u_\ell - u_\infty)||_{2,\Omega_{\ell_1/2}} \leq \frac{C}{\ell_1}||\nabla(u_\ell - u_\infty)||_{2,\Omega_{\ell_1}} \qquad (7.23)$$

for any $\ell_1 \le \ell$. Taking $\ell_1 = \ell/2^{k-1}$ and iterating the formula (7.23) we get

$$||\nabla(u_\ell - u_\infty)||_{2,\Omega_{\ell/2^k}} \le \frac{C}{\ell^k}||\nabla(u_\ell - u_\infty)||_{2,\Omega_\ell}$$

where C is a constant independent of ℓ. It follows that

$$||\nabla(u_\ell - u_\infty)||_{2,\Omega_{\ell/2^k}} \le \frac{C}{\ell^k}\{||\nabla u_\ell||_{2,\Omega_\ell} + ||\nabla u_\infty||_{2,\Omega_\ell}\}. \qquad (7.24)$$

Since u_∞ depends on x_2 only we have

$$||\nabla u_\infty||_{2,\Omega_\ell} \le (2\ell)^{p/2}||\nabla u_\infty||_{2,\omega}. \qquad (7.25)$$

Recalling (7.11) we derive from (7.24) that for some constant C independent of ℓ it holds that

$$||\nabla(u_\ell - u_\infty)||_{2,\Omega_{\ell/2^k}} \le \frac{C}{\ell^{k-p/2}}.$$

Choosing then k such that $k - \frac{p}{2} \ge r$, ℓ such that $\frac{\ell}{2^k} \ge \ell_0$ we obtain

$$||\nabla(u_\ell - u_\infty)||_{2,\Omega_{\ell_0}} \le \frac{C}{\ell^r}$$

which completes the proof of Theorem 5.1. $\qquad\qquad\square$

7.2. The system of elasticity

In this section we assume that $n = N = 3$ and we consider a physical issue, namely the system of elasticity (see Section 6.2). Let ω be a bounded open subset of $\mathbb{R}^2$ with a C^1 or eventually Lipschitz continuous boundary (see Remark (6.1)). We suppose that our reference configuration is the cylindrical domain

$$\Omega_\ell = (-\ell, \ell) \times \omega \qquad (7.26)$$

i.e., Ω_ℓ is the domain occupied by the elastic body that we consider in its un-deformed state (see Figure 7.1). We suppose that this body is loaded with some

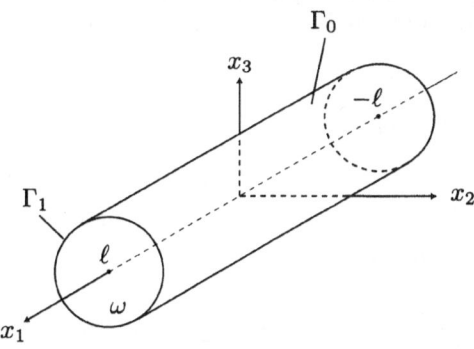

FIGURE 7.1

internal forces of density

$$f = (0, f_2(x_2, x_3), f_3(x_2, x_3)) \tag{7.27}$$

i.e., depending only on the transversal coordinates, and also directed transversally, and with some superficial forces on $\Gamma_1 = \{\ell\} \times \omega$

$$g(\ell, x_2, x_3) = (g_1(\ell, x_2, x_3), g_2(\ell, x_2, x_3), g_3(\ell, x_2, x_3)). \tag{7.28}$$

In other words with our usual notation $x = (X_1, X_2)$ where here $X_1 = x_1$, $X_2 = (x_2, x_3)$ we set

$$f = (0, f_2(X_2), f_3(X_2)), \qquad g = (g_1(\ell, X_2), g_2(\ell, X_2), g_3(\ell, X_3)). \tag{7.29}$$

To simplify we suppose that

$$f \in \mathbb{L}^2(\omega), \qquad g \in \mathbb{L}^2(\Gamma_1). \tag{7.30}$$

(Recall that $\mathbb{L}^2(\omega) = (L^2(\omega))^3$, $\mathbb{L}^2(\Gamma_1) = (L^2(\Gamma_1))^3$.)

Let $\lambda \geq 0$, $\mu > 0$ be two constants. Then, due to Theorems 6.17, 6.20 there exists $u_\infty = (u_\infty^2(X_2), u_\infty^3(X_2))$ unique solution to

$$\begin{cases} u_\infty \in \mathbb{H}_0^1(\omega) = (H_0^1(\omega))^2, \\ \lambda \int_\omega \operatorname{div}_2 u_\infty \operatorname{div}_2 v \, dX_2 + 2\mu \sum_{i,j=2}^3 \int_\omega \varepsilon_{ij}(u_\infty)\varepsilon_{ij}(v) \, dX_2 = \int_\omega fv \\ \forall v = (v^2, v^3) \in \mathbb{H}_0^1(\omega). \end{cases} \tag{7.31}$$

Recall that for $v = (v^2, v^3)$

$$\varepsilon_{ij}(v) = \frac{1}{2}(\partial_{x_j} v^i + \partial_{x_i} v^j) \quad i, j = 2, 3,$$

and in the above integrals we have denoted by div_2 the operator defined by

$$\operatorname{div}_2 v = \partial_{x_2} v^2 + \partial_{x_3} v^3 \quad \forall v = (v^2, v^3). \tag{7.32}$$

The function u_∞ represents the displacement of a two-dimensional body ω maintained fixed on its boundary and loaded by a density of forces f (see Chapter 6). We define then

$$\Gamma_0 = (-\ell, \ell) \times \partial\omega \cup \{-\ell\} \times \omega \tag{7.33}$$

where $\partial\omega$ is the boundary of ω and

$$\mathbb{H}_0^1(\Omega_\ell; \Gamma_0) = (H_0^1(\Omega_\ell; \Gamma_0))^3. \tag{7.34}$$

Then, by Theorem 6.20, there exists a unique u_ℓ solution to

$$\begin{cases} u_\ell \in \mathbb{H}_0^1(\Omega_\ell; \Gamma_0), \\ \lambda \int_{\Omega_\ell} \operatorname{div} u_\ell \operatorname{div} v \, dx + 2\mu \sum_{i,j=1}^3 \int_{\Omega_\ell} \varepsilon_{ij}(u_\ell)\varepsilon_{ij}(v) \, dx \\ = \int_{\Omega_\ell} fv \, dx + \int_{\Gamma_1} gv \, d\gamma(x) \quad \forall v \in \mathbb{H}_0^1(\Omega_\ell; \Gamma_0). \end{cases} \tag{7.35}$$

Here, as in Chapter 6, the symbol div denotes the usual divergence operator. It is clear that u_ℓ is the displacement of a cylindrical body clamped along Γ_0 and loaded by forces of density f and g respectively inside Ω and on Γ_1. If we set

$$u_\infty = (0, u_\infty^2, u_\infty^3) \tag{7.36}$$

then we have:

THEOREM 7.3. *Under the above assumptions, $\forall \ell_0 > 0$, $\forall r > 0$ there exists a constant C independent of ℓ such that*

$$||\nabla(u_\ell - u_\infty)||_{2,\Omega_{\ell_0}} \leq \frac{C}{\ell^r}. \tag{7.37}$$

In particular $u_\ell \to u_\infty$ in $\mathbb{H}^1(\Omega_{\ell_0}) = (H^1(\Omega_{\ell_0}))^3$.

As usual we will need an estimate on u_ℓ. For this we have

LEMMA 7.4. *Let u_ℓ be the solution to (7.35). There exists a constant C independent of ℓ such that*

$$||\nabla u_\ell||_{2,\Omega_\ell} \leq C\ell^{1/2}\{||f||_{2,\omega} + ||g||_{2,\Gamma_1}\}. \tag{7.38}$$

PROOF. We choose $v = u_\ell$ in (7.35). We get

$$\lambda|\operatorname{div} u_\ell|_{2,\Omega_\ell}^2 + 2\mu||\varepsilon(u_\ell)||_{2,\Omega_\ell}^2 = \int_{\Omega_\ell} f u_\ell \, dx + \int_{\Gamma_1} g u_\ell \, d\gamma(x). \tag{7.39}$$

Noting that $u_\ell(\ell - x_1, x_2, x_3) \in \mathbb{H}^1(\mathbb{R}_+^3)$ and has a compact support, we derive easily from (6.32) that for some universal constant

$$||\nabla u_\ell||_{2,\Omega_\ell}^2 \leq C||\varepsilon(u_\ell)||_{2,\Omega_\ell}^2 \leq C\{||f||_{2,\Omega_\ell}||u_\ell||_{2,\Omega_\ell} + ||g||_{2,\Gamma_1}||u_\ell||_{2,\Gamma_1}\}. \tag{7.40}$$

(Using for instance (3.28), see also Remark 3.2.) It is clear that for some constant C independent of ℓ,

$$||u_\ell||_{2,\Omega_\ell} \leq C||\nabla u_\ell||_{2,\Omega_\ell}. \tag{7.41}$$

Moreover for u smooth in $H^1(\Omega_\ell; \Gamma_0)$,

$$u(\ell, x_2, x_3) = \int_{-\ell}^{\ell} \partial_{x_1} u(s, x_2, x_3) \, ds \leq \left\{ \int_{-\ell}^{\ell} \partial_{x_1} u(s, x_2, x_3)^2 \, ds \right\}^{1/2} \sqrt{2\ell}.$$

(We applied the Cauchy–Schwarz inequality.) Squaring and integrating in $\mathrm{X}_2 = (x_2, x_3)$ we obtain

$$|u|_{2,\Gamma_1}^2 = \int_\omega u^2(\ell, \mathrm{X}_2) \, d\mathrm{X}_2 \leq 2\ell|\partial_{x_1} u|_{2,\Omega_\ell}^2. \tag{7.42}$$

Hence we have for u_ℓ

$$||u_\ell||_{2,\Gamma_1} \leq C\ell^{1/2}||\nabla u_\ell||_{2,\Omega_\ell}. \tag{7.43}$$

Moreover, it is clear that

$$||f||_{2,\Omega_\ell}^2 = \int_{(-\ell,\ell)} \int_\omega |f|^2 \, d\mathrm{X}_2 \, dx_1 = 2\ell||f||_{2,\omega}^2. \tag{7.44}$$

Thus, combining (7.40)–(7.44) we obtain for some constant C independent of ℓ

$$||\nabla u_\ell||^2_{2,\Omega_\ell} \le C\{||f||_{2,\omega} + ||g||_{2,\Gamma_1}\}\ell^{1/2}||\nabla u_\ell||_{2,\Omega_\ell}$$

and (7.38) follows. This completes the proof of the lemma. $\square$

We now turn to the proof of the theorem.

PROOF OF THEOREM 7.3. We consider a smooth function ϱ as for instance in Figure 2.2. Then, clearly, for any $\ell_1 \le \ell$

$$\varrho^2\left(\frac{x_1}{\ell_1}\right)(u_\ell - u_\infty) \in \mathbb{H}^1_0(\Omega_\ell). \tag{7.45}$$

Thus, from (7.35), (7.31) we derive

$$
\lambda \int_{\Omega_\ell} \operatorname{div} u_\ell \operatorname{div}\{\varrho^2(u_\ell - u_\infty)\} \, dx
$$

$$
+ 2\mu \sum_{i,j=1}^3 \int_{\Omega_\ell} \varepsilon_{ij}(u_\ell)\varepsilon_{ij}\{\varrho^2(u_\ell - u_\infty)\} \, dx
$$

$$
= \int_{(-\ell,\ell)} \int_\omega f\{\varrho^2(u_\ell - u_\infty)\} \, dx \tag{7.46}
$$

$$
= \lambda \int_{\Omega_\ell} \operatorname{div}_2 u_\infty \operatorname{div}_2\{\varrho^2(u_\ell - u_\infty)\} \, dx
$$

$$
+ 2\mu \sum_{i,j=2}^3 \int_{\Omega_\ell} \varepsilon_{ij}(u_\ell)\varepsilon_{ij}\{\varrho^2(u_\ell - u_\infty)\} \, dx
$$

where we have set $\varrho^2 = \varrho^2(\frac{x_1}{\ell_1})$. First, remark that due to (7.36) we have

$$\operatorname{div}_2 u_\infty = \operatorname{div} u_\infty.$$

Moreover – since $\operatorname{div}_2 u_\infty$ is independent of x_1 – it holds that

$$\int_{\Omega_\ell} \operatorname{div}_2 u_\infty \partial_{x_1}\{\varrho^2(u_\ell^1 - u_\infty^1)\} \, dx = 0.$$

Thus,

$$\int_{\Omega_\ell} \operatorname{div}_2 u_\infty \operatorname{div}_2\{\varrho^2(u_\ell - u_\infty)\} \, dx = \int_{\Omega_\ell} \operatorname{div} u_\infty \operatorname{div}\{\varrho^2(u_\ell - u_\infty)\} \, dx. \tag{7.47}$$

Next for $j = 1, 2, 3$ it holds that

$$\varepsilon_{1j}(u_\infty) = \varepsilon_{j1}(u_\infty) = \frac{1}{2}\{\partial_{x_j} u_\infty^1 + \partial_{x_1} u_\infty^j\} = 0. \tag{7.48}$$

Thus from (7.46) we obtain

$$\lambda \int_{\Omega_\ell} \operatorname{div}(u_\ell - u_\infty) \operatorname{div}\{\varrho^2(u_\ell - u_\infty)\} \, dx$$

$$+ 2\mu \int_{\Omega_\ell} \varepsilon_{ij}(u_\ell - u_\infty)\varepsilon_{ij}\{\varrho^2(u_\ell - u_\infty)\} \, dx = 0 \tag{7.49}$$

where the summation in i, j is carried out on all indices 1,2,3.

We set $w = u_\ell - u_\infty$. It holds that

$$\varepsilon_{ij}(\varrho^2 w) = \frac{1}{2}\{\partial_{x_i}(\varrho\varrho w^j) + \partial_{x_j}(\varrho\varrho w^j)\}$$

$$= \frac{1}{2}\varrho\{\partial_{x_i}(\varrho w^j) + \partial_{x_j}(\varrho w^i)\} + \frac{1}{2}\varrho(\partial_{x_i}\varrho w^j + \partial_{x_j}\varrho w^i) \tag{7.50}$$

$$= \varrho\varepsilon_{ij}(\varrho w) + \varrho\frac{1}{2}\{\partial_{x_i}\varrho w^j + \partial_{x_j}\varrho w^i\},$$

$$\varrho\varepsilon_{ij}(w) = \varepsilon_{ij}(\varrho w) - \frac{1}{2}\{\partial_{x_i}\varrho w^j + \partial_{x_j}\varrho w^i\}. \quad \cdot \tag{7.51}$$

We deduce that

$$\varepsilon_{ij}(w)\varepsilon_{ij}(\varrho^2 w) = \varrho\varepsilon_{ij}(w)\varepsilon_{ij}(\varrho w) + \frac{1}{2}\varrho\varepsilon_{ij}(w)\{\partial_{x_i}\varrho w^j + \partial_{x_j}\varrho w^i\}$$

$$= \varepsilon_{ij}(\varrho w)\varepsilon_{ij}(\varrho w) - \frac{1}{2}\varepsilon_{ij}(\varrho w)\{\partial_{x_i}\varrho w^j + \partial_{x_j}\varrho w^i\}$$

$$+ \frac{1}{2}\varepsilon_{ij}(\varrho w)\{\partial_{x_i}\varrho w^j + \partial_{x_j}\varrho w^i\} \tag{7.52}$$

$$- \frac{1}{4}\{\partial_{x_i}\varrho w^j + \partial_{x_j}\varrho w^i\}^2$$

$$= \varepsilon_{ij}(\varrho w)^2 - \frac{1}{4}\{\partial_{x_i}\varrho w^j + \partial_{x_j}\varrho w^i\}^2.$$

Similarly we have with the summation convention

$$\operatorname{div}(\varrho^2 w) = \partial_{x_i}(\varrho^2 w^i) = \varrho\partial_{x_i}(\varrho w^i) + \varrho\nabla\varrho \cdot w = \varrho\operatorname{div}(\varrho w) + \varrho\nabla\varrho \cdot w,$$

$$\varrho\operatorname{div}(w) = \operatorname{div}(\varrho w) - \nabla\varrho \cdot w.$$

Hence it follows that

$$\operatorname{div} w \operatorname{div}(\varrho^2 w) = \varrho\operatorname{div} w \operatorname{div}(\varrho w) + \varrho\operatorname{div} w\nabla\varrho \cdot w = \operatorname{div}(\varrho w)^2 - (\nabla\varrho \cdot w)^2. \tag{7.53}$$

Using (7.52), (7.53) in (7.49) we derive, with a summation in i, j,

$$\lambda \int_{\Omega_{\ell_1}} \operatorname{div}(\varrho w)^2 \, dx + 2\mu \int_{\Omega_{\ell_1}} \varepsilon_{ij}^2(\varrho w) \, dx$$

$$= \lambda \int_{\Omega_{\ell_1}} (\nabla\varrho \cdot w)^2 \, dx + 2\mu \int_{\Omega_{\ell_1}} \frac{1}{4}\{\partial_{x_i}\varrho w^j + \partial_{x_j}\varrho w^i\}^2 \, dx.$$

(Note that it is enough to integrate on Ω_{ℓ_1}.) Noting that

$$\nabla\varrho = \left(\frac{1}{\ell_1}\partial_{x_1}\varrho\left(\frac{x_1}{\ell_1}\right), 0, 0\right)$$

and recalling the inequality (6.27) we derive for some constant C independent of $\ell_1 \leq \ell$ that

$$\int_{\Omega_{\ell_1}} |\nabla(\varrho w)|^2 \, dx \leq \frac{C}{\ell_1^2} \int_{\Omega_{\ell_1}} |w|^2 \, dx.$$

Recalling the definition of w and ϱ it follows that

$$\int_{\Omega_{\ell_1/2}} |\nabla(u_\ell - u_\infty)|^2 \, dx \leq \frac{C}{\ell_1^2} \int_{\Omega_{\ell_1}} |u_\ell - u_\infty|^2 \, dx.$$

From (3.29) and Remark 3.2 we deduce that

$$\left\| \nabla(u_\ell - u_\infty) \right\|_{2,\Omega_{\ell_1/2}} \leq \frac{C}{\ell_1} \left\| \nabla(u_\ell - u_\infty) \right\|_{2,\Omega_{\ell_1}} \quad \forall \, \ell_1 \leq \ell.$$

By iteration of this formula we get

$$\left\| \nabla(u_\ell - u_\infty) \right\|_{2,\Omega_{\ell/2^k}} \leq \frac{C}{\ell^k} \left\| \nabla(u_\ell - u_\infty) \right\|_{2,\Omega_\ell} \tag{7.54}$$

where C is independent of ℓ. It is clear that

$$\left\| \nabla u_\infty \right\|_{2,\Omega_\ell} = \sqrt{2}\ell^{1/2} \left\| \nabla u_\infty \right\|_{2,\omega}$$

and thus by (7.38) and (7.54) we obtain

$$\left\| \nabla(u_\ell - u_\infty) \right\|_{2,\Omega_{\ell/2^k}} \leq \frac{C}{\ell^k} \left\{ \left\| \nabla u_\ell \right\|_{2,\Omega_\ell} + \left\| \nabla u_\infty \right\|_{2,\Omega_\ell} \right\} \leq \frac{C}{\ell^{k-1/2}}$$

where C is independent of ℓ. Choosing as usual

$$k - \frac{1}{2} \geq r, \qquad \ell/2^k \geq \ell_0$$

the result follows. This completes the proof of the theorem. $\qquad\qquad\square$

REMARK 7.1. Note that the limit of u_ℓ is completely independent of the force g acting on Γ_1. Note also that the first component u_ℓ^1 is forced to converge toward 0 when $\ell \to +\infty$. The above technique allows us to extend also Theorem 7.1 in the case of constant coefficients satisfying the Legendre–Hadamard condition.

Open problems

1. It would be interesting to see when an exponential rate of convergence is available in Theorem 7.1.
2. Convergence in higher order Sobolev spaces should be established, especially in the case of Theorem 7.3.
3. Is an exponential rate of convergence true in Theorem 7.3?
4. Some generalizations of Theorem 7.3 should be established (for instance for the pure traction problem).
5. Nothing is known regarding the asymptotic behaviour of nonlinear systems.
6. Convergence results for the Stokes problem are not known.

Chapter 8

Parabolic Equations

The problems we shall solve now are of the so-called heat equation type. Such theory is available in many places (see for instance [21], [2], [6], [17], [8]) but for the self-completeness of this book we would like to recall the main ideas. Ω being a domain of $\mathbb{R}^n$, we would like, for instance, to find a function $u(x,t)$ such that

$$\begin{cases} u_t - \Delta u = f & \text{in } \Omega \times (0,T), \\ u(x,t) = 0 & \text{on } \Gamma \times (0,T), \\ u(x,0) = u_0(x) & \text{on } \Omega, \end{cases} \qquad (8.1)$$

$f(x,t)$, $u_0(x)$ are two given data. Clearly, the second equation of (8.1) can also be interpreted as

$$u(\cdot,t) \in H_0^1(\Omega). \qquad (8.2)$$

Moreover – assuming that all the integrations that we are performing make sense – if for $v \in H_0^1(\Omega)$ we multiply the first equation of (8.1) by v and integrate over Ω we get

$$\int_\Omega u_t v \, dx + \int_\Omega \nabla u \nabla v \, dx = \int_\Omega f v \, dx \quad \forall v \in H_0^1(\Omega).$$

Introducing

$$(u,v) = \int_\Omega uv \, dx, \quad ((u,v)) = \int_\Omega \nabla u \nabla v \, dx,$$

the above equality reads

$$\frac{d}{dt}(u,v) + ((u,v)) = (f,v) \quad \forall v \in H_0^1(\Omega). \qquad (8.3)$$

So, we can weaken our initial problem and look for u such that (8.3) holds for instance in $\mathcal{D}'(0,T)$.

8.1. Functional spaces for parabolic problems

Let us denote by X a Banach space, with norm $\| \cdot \|_X$.

DEFINITION 8.1. For $a,b \in \mathbb{R}$ we denote by

$$L^p(a,b;X), \quad 1 \le p < +\infty \qquad (8.4)$$

the space of (class of functions) $f : (a,b) \to X$ that are measurable and such that

$$\int_a^b \|f(t)\|_X^p \, dt < +\infty, \qquad (8.5)$$

and by $L^\infty(a, b; X)$ the space of functions essentially bounded on (a, b) – i.e., such that there exists M such that

$$\|f(t)\|_X \leq M \quad \text{a.e. } t \in (a, b). \tag{8.6}$$

Then it is easy to show

THEOREM 8.1. *Equipped with the norms*

$$\|f\|_{L^p(a,b;X)} = \left(\int_a^b \|f(t)\|_X^p \, dt \right)^{1/p}, \quad 1 \leq p < +\infty,$$

$$\|f\|_{L^\infty(a,b;X)} = \text{Inf}\{ M \in \mathbb{R} \mid \|f(t)\|_X \leq M \text{ a.e. } t \in (a, b) \},$$

the spaces $L^p(a, b; X)$, $1 \leq p \leq +\infty$ *are Banach spaces.*

REMARK 8.1. For every $f \in L^p(a, b; X)$ we can define

$$\int_a^b f(t) \, dt \in X$$

(see [5]).

Let $-\infty \leq a < b \leq +\infty$.

DEFINITION 8.2. A vector valued distribution on (a, b) with values in X is a linear mapping

$$\mathcal{D}(a, b) \rightarrow X,$$

$$\varphi \mapsto \langle T, \varphi \rangle,$$

continuous on $\mathcal{D}(a, b)$ in the sense that

$$\lim_{i \to +\infty} \langle T, \varphi_i \rangle = \langle T, \varphi \rangle \quad \text{in } X$$

for any sequence φ_i such that $\varphi_i \to \varphi$ in $\mathcal{D}(a, b)$ (see [33], [8]).

EXAMPLE. Let $u \in L^1_{\text{loc}}(a, b; X)$; then

$$\varphi \mapsto \int_a^b u(s)\varphi(s) \, ds$$

is a distribution on (a, b) with values in X. We say that this is the distribution defined by u. Two distinct functions define the same distribution iff they agree almost everywhere.

NOTATION. We will denote by $\mathcal{D}'(a, b; X)$ the space of distributions on (a, b) with values in X.

DEFINITION 8.3. Let $T \in \mathcal{D}'(a, b; X)$. Then for every integer k we define a distribution $T^{(k)}$ by the formula

$$\langle T^{(k)}, \varphi \rangle = (-1)^k \left\langle T, \frac{d^k \varphi}{dt^k} \right\rangle. \tag{8.7}$$

$T^{(k)}$ is called the k-th derivative of T on (a, b) and will also be denoted

$$\frac{d^k T}{dt^k} = T^{(k)}.$$

EXAMPLES. Let $u \in L^1(a, b; X)$. Then $u^{(1)}$ will also be denoted u' or u_t if t is the variable in (a, b). If $X = L^1(\Omega)$ for some domain Ω of $\mathbb{R}^n$, then $u \in L^1(a, b; L^1(\Omega))$ can be identified with a function of two variables by the formula

$$u(t)(x) = u(x, t).$$

We can easily show that, for smooth functions,

$$u_t(t)(x) = \frac{\partial u}{\partial t}(x, t),$$

thus we shall use both notations.

REMARK 8.2. If X, Y are two Banach spaces such that

$$X \hookrightarrow Y \quad \text{(continuous embedding)}, \tag{8.8}$$

then clearly

$$\mathcal{D}'(a, b; X) \hookrightarrow \mathcal{D}'(a, b; Y)$$

and

$$L^p(a, b; X) \hookrightarrow L^p(a, b; Y) \quad \forall 1 \leq p \leq +\infty.$$

DEFINITION 8.4. Let X, Y be as in (8.8). Let $u \in L^1(a, b; X)$. We say that $u' \in L^p(a, b; Y)$ iff there exists $v \in L^p(a, b; Y)$ such that

$$-\int_a^b u(t)\varphi'(t)\, dt = \int_a^b v(t)\varphi(t)\, dt \quad \forall \varphi \in \mathcal{D}(a, b) \tag{8.9}$$

(this equality takes place in Y).

In what follows we will have to consider the situation of two Hilbert spaces V, H such that

$$V \hookrightarrow H \hookrightarrow V', \tag{8.10}$$
$$V \text{ dense in } H, \tag{8.11}$$

where V' is the dual of V. We will denote by

$((\,,\,))$, $\|\ \|$ respectively the scalar product and the norm in V,

$(\,,\,)$, $|\ |$ respectively the scalar product and the norm in H.

A suitable choice could be:

$$H_0^1(\Omega) \subset L^2(\Omega) \subset H^{-1}(\Omega).$$

Then we can define

$$H^1(a, b; V, V') = \{\, u \in L^2(a, b; V) \,|\, u_t \in L^2(a, b; V') \,\}. \tag{8.12}$$

To define a Hilbertian norm in V' by the Riesz representation theorem, we identify V' and V through the formula

$$\langle f, v \rangle_{V',V} = ((\tilde{f}, v))$$

and we set

$$|f|_{V'} = \|\tilde{f}\| = |\tilde{f}|_V.$$

Of course this norm is equivalent to the norm of strong dual.

Now we can state

THEOREM 8.2. $H^1(a, b; V, V')$ is a Hilbert space for the norm

$$\|u\|_1^2 = \|u\|_{L^2(a,b;V)}^2 + \|u_t\|_{L^2(a,b;V')}^2. \tag{8.13}$$

PROOF. See for instance [17] or [8].   □

Let us denote by $\mathcal{D}([a, b]; V)$ the space of restrictions to $[a, b]$ of functions of $\mathcal{D}(\mathbb{R}; V)$. ($\mathcal{D}(\mathbb{R}; V)$ is the space of functions indefinitely differentiable with compact support and values in V). Therefore we can proceed to

THEOREM 8.3. $\mathcal{D}([a, b]; V)$ is dense in $H^1(a, b; V, V')$.

PROOF. See for instance [17] or [8].   □

Now we will need the following lemma:

LEMMA 8.4. There exists a continuous linear operator of extension that we denote by P, from $H^1(a, b; V, V')$ into $H^1(-\infty, +\infty; V, V')$ such that

$$Pu = u \quad a.e. \ on \ (a, b) \quad \forall u \in H^1(a, b; V, V').$$

PROOF. We can always assume b finite. If $a = -\infty$ then one sets

$$Pu(t) = \begin{cases} u(t) & t \leq b, \\ u(2b - t) & t \geq b. \end{cases}$$

It is easy to show that this solves the problem. If a is finite, then setting

$$u = \theta_1 u + \theta_2 u$$

where θ_1, θ_2 are two smooth functions such that

$$\theta_1 = 0 \ \text{on} \ (-\infty, a], \quad \theta_2 = 0 \ \text{on} \ [b, +\infty), \quad \theta_1 + \theta_2 = 1 \ \text{on} \ [a, b],$$

we have clearly – recall (8.13):

$$\|\theta_1 u\|_1 \leq C\|u\|_1,$$
$$\|\theta_2 u\|_1 \leq C\|u\|_1,$$

and we conclude using the previous step.   □

Then we can show

THEOREM 8.5. *Let $u \in H^1(a, b; V, V')$. Then u can be identified with a continuous function on $[a, b]$ with values in H. Moreover*

$$H^1(a, b; V, V') \hookrightarrow C([a, b]; H) \tag{8.14}$$

where $C([a, b]; H)$ denotes the space of continuous functions on $[a, b]$ with values in H equipped with the topology of the uniform convergence on $[a, b]$.

PROOF. Let $u \in H^1(a, b; V, V')$ and $Pu \in H^1(-\infty, +\infty; V, V')$ the extension introduced in Lemma 8.4. Let $\varphi_n \in \mathcal{D}(\mathbb{R}, V)$ such that

$$\varphi_n \to Pu \quad \text{in } H^1(-\infty, +\infty; V, V').$$

Then we have:

$$\frac{1}{2}\frac{d}{dt}|\varphi_n(t)|^2 = \left(\frac{d\varphi_n}{dt}, \varphi_n\right) = \left\langle \frac{d\varphi_n}{dt}, \varphi_n \right\rangle \le \|\varphi_n'\|\|\varphi_n\|$$

(the duality in the bracket is the V'-V duality; note that for two vectors of V, $V \hookrightarrow H \hookrightarrow V'$ we have $(u, v) = \langle u, v \rangle$, by identifying u with the linear form $v \mapsto (u, v)$ that it defines on V in V'). Integrating we obtain

$$|\varphi_n(t)|^2 \le 2 \int_{-\infty}^{t} \|\varphi_n'\|\|\varphi_n\|\, dt.$$

Using the Young inequality

$$ab \le \frac{a^2}{2} + \frac{b^2}{2}$$

it follows that

$$|\varphi_n(t)|^2 \le \|\varphi_n'\|_{L^2(-\infty,+\infty;V')}^2 + \|\varphi_n\|_{L^2(-\infty,+\infty;V)}^2,$$

i.e.,

$$|\varphi_n(t)|^2 \le \|\varphi_n\|_{H^1(-\infty,+\infty;V,V')}^2 \quad \forall t \in \mathbb{R}. \tag{8.15}$$

Thus if we apply this inequality to $\varphi_n - \varphi_m$, we see that φ_n is a Cauchy sequence in $C((-\infty, +\infty), H)$ so converges uniformly in this space. This implies that $Pu \in C(-\infty, +\infty; H)$ – thus $u \in C([a, b], H)$ and passing to the limit in (8.15) we get

$$|u|_{C([a,b],H)} \le \|Pu\|_{H^1(-\infty,+\infty;V,V')} \le C\|u\|_{H^1(a,b;V,V')}$$

which gives (8.14). $\qquad\square$

A fundamental consequence of Theorem 8.5 is that it allows us to define for $u \in H^1(a, b; V, V')$ the trace $u(a), u(b)$ – i.e., these values are the values in H of the function $u \in C([a, b]; H)$. From that argument we can deduce the Green formula:

THEOREM 8.6. *Let $u, v \in H^1(a, b; V, V')$, a, b finite; then*

$$\int_a^b \langle u'(t), v(t) \rangle\, dt + \int_a^b \langle v'(t), u(t) \rangle\, dt = (u(b), v(b)) - (u(a), v(a)). \tag{8.16}$$

PROOF. For $u, v \in \mathcal{D}([a, b], V)$ we have

$$\int_a^b \langle u'(t), v(t) \rangle \, dt + \int_a^b \langle v'(t), u(t) \rangle \, dt = \int_a^b \frac{d}{dt} (u(t), v(t)) = u(b)v(b) - u(a)v(a).$$

Then the result follows by density. $\square$

We can also state

THEOREM 8.7. *If $u \in H^1(a, b; V, V')$, then for all $v \in V$*

$$\frac{d}{dt}(u(\cdot), v) = \langle u_t(\cdot), v \rangle \quad \text{in } \mathcal{D}'(a, b). \tag{8.17}$$

PROOF. Let $\varphi \in \mathcal{D}(a, b)$. From (8.16) we derive

$$\int_a^b \langle u'(t), \varphi(t)v \rangle + \langle \varphi'(t)v, u(t) \rangle \, dt = 0 \quad \forall v \in V.$$

This can also be written

$$\int_a^b \langle u'(t), v \rangle \varphi(t) \, dt = - \int_a^b \langle v, u(t) \rangle \varphi'(t) \, dt = - \int_a^b (u(t), v) \varphi'(t) \, dt,$$

hence the result. $\square$

8.2. Linear parabolic problems

Let us analyze the setting of (8.1)–(8.3). We shall use the three spaces

$$H_0^1(\Omega) \subset L^2(\Omega) \subset H^{-1}(\Omega).$$

Indeed, one could choose f in $H^{-1}(\Omega)$. Moreover we also consider a bilinear form $((\,,\,))$ which by multiplying it by a function of t, we could let depend on t. So, let us consider the following abstract situation already partly seen in the previous section.

Let V, H be two real Hilbert spaces with the following properties:

$$V, H \text{ are separable}, \tag{8.18}$$

$$V \hookrightarrow H \hookrightarrow V', \ V \text{ dense in } H. \tag{8.19}$$

(Recall that $\hookrightarrow$ means that the canonical embedding is continuous.) For the norm and scalar product we will copy the H_0^1-L^2 situation, in other words:

- $((\cdot, \cdot))$, $\| \cdot \|$ will denote the scalar product and the norm in V,
- $(\cdot, \cdot)$, $| \cdot |$ will denote the scalar product and the norm in H.

Then, for $t \in (0, T)$ we denote by $a(t; u, v)$ a bilinear form on V with the following properties:

$$\forall u, v \in V, \ t \mapsto a(t; u, v) \text{ is measurable}, \tag{8.20}$$

$\exists M$ a constant such that

$$|a(t; u, v)| \leq M \|u\| \|v\| \quad \forall u, v \in V, \text{ a.e. } t \in (0, T), \tag{8.21}$$

$\exists$ two positive constants α, λ such that

$$a(t; u, u) + \lambda |u|^2 \geq \alpha \|u\|^2 \quad \forall u \in V, \text{ a.e. } t \in (0, T). \tag{8.22}$$

The assumption (8.21) is a hypothesis of continuity, the assumption (8.22) a hypothesis of coerciveness.

Then for

$$u_0 \in H, \quad f \in L^2(0, T; V') \tag{8.23}$$

we would like to find a function u solution to the problem

$$u \in L^2(0, T; V), \quad u_t \in L^2(0, T; V'), \tag{8.24}$$

$$u(0) = u_0, \tag{8.25}$$

$$\frac{d}{dt}(u, v) + a(\cdot; u, v) = \langle f, v \rangle \text{ in } \mathcal{D}'(0, T) \quad \forall v \in V. \tag{8.26}$$

Note that (8.25) makes sense by (8.14). Also, since (see (8.17))

$$\frac{d}{dt}(u, v) = \langle u_t, v \rangle$$

the equation (8.26) can be written

$$u_t + a(t; u, \cdot) = f \quad \text{in } L^2(0, T; V'). \tag{8.27}$$

Then we can prove

THEOREM 8.8. *Under the above assumptions (8.18)–(8.22) there exists a unique solution to (8.24)–(8.26).*

PROOF. See for instance [**17**] or [**8**]. $\qquad\square$

We turn now to the applications. Let Ω be a bounded open set of $\mathbb{R}^n$ with boundary Γ. Let

$$H = L^2(\Omega) \tag{8.28}$$

and V be a closed subspace of $H^1(\Omega)$ such that

$$H_0^1(\Omega) \subset V \subset H^1(\Omega). \tag{8.29}$$

Let a_{ij}, a_i, $i, j = 1, \ldots, n$, a_0 be functions of $L^\infty(\Omega \times (0, T))$ such that

$$a_{ij}(x, t)\xi_i\xi_j \geq \alpha |\xi|^2 \quad \text{a.e. } (x, t) \in \Omega \times (0, T), \quad \forall \xi \in \mathbb{R}^n. \tag{8.30}$$

Then we can set for $u, v \in V$

$$a(t; u, v) = \int_\Omega \left\{ a_{ij} \frac{\partial u}{\partial x_j} \frac{\partial v}{\partial x_i} + a_i \frac{\partial u}{\partial x_i} v + a_0 u v \right\} dx. \tag{8.31}$$

The hypothesis (8.21) is easy to check. Moreover, see (8.30),

$$a(t; u, u) = \int_\Omega \left\{ a_{ij} \frac{\partial u}{\partial x_j} \frac{\partial u}{\partial x_i} + a_i \frac{\partial u}{\partial x_i} u + a_0 u^2 \right\} dx$$

$$\geq \alpha \int_\Omega |\nabla u|^2 \, dx - \|a\|_\infty \int_\Omega |\nabla u| \, |u| \, dx - |a_0|_\infty \int_\Omega u^2 \, dx$$

where $|\cdot|_\infty$ denotes the $L^\infty(\Omega)$ norm and $|\cdot|$ the euclidean norm, a the vector with entries a_i. Using Young's inequality we get

$$a(t; u, u) \geq \alpha \int_\Omega |\nabla u|^2 \, dx - ||a||_\infty \frac{\varepsilon}{2} \int_\Omega |\nabla u|^2 \, dx$$

$$- ||a||_\infty \frac{1}{2\varepsilon} \int_\Omega u^2 \, dx - |a_0|_\infty \int_\Omega u^2 \, dx,$$

i.e.,

$$a(t; u, u) + \left(\frac{||a||_\infty}{2\varepsilon} + |a_0|_\infty \right) \int_\Omega u^2 \, dx \geq \left(\alpha - \frac{\varepsilon ||a||_\infty}{2} \right) \int_\Omega |\nabla u|^2 \, dx.$$

Setting

$$\lambda = \frac{||a||_\infty}{2\varepsilon} + |a_0|_\infty + \alpha - \frac{\varepsilon ||a||_\infty}{2}$$

this inequality can also be written, for a.e. $t \in (0, T)$,

$$a(t; u, u) + \lambda \int_\Omega u^2 \, dx \geq \left(\alpha - \frac{\varepsilon ||a||_\infty}{2} \right) \int_\Omega \{ |\nabla u|^2 + u^2 \} \, dx$$

which is exactly (8.22) with α replaced by $\alpha - \frac{\varepsilon ||a||_\infty}{2} \geq \frac{\alpha}{2}$ for ε small enough. Taking

$$u_0 \in L^2(\Omega), \qquad f \in L^2(0, T; V'), \tag{8.32}$$

Theorem 8.8 leads directly to

THEOREM 8.9. *Under the above assumptions there exists a unique solution u to*

$$\begin{cases} u \in L^2(0, T; V), \quad u_t \in L^2(0, T; V'), \\ u(0) = u_0, \\ \dfrac{d}{dt}(u, v) + a(\cdot; u, v) = \langle f, v \rangle \text{ in } \mathcal{D}'(0, T), \quad \forall v \in V. \end{cases} \tag{8.33}$$

Let us interpret the problem (8.33). More precisely, let us assume that all the functions at stake are smooth in such a way that all the following computations make sense. The last equation of (8.33) reads

$$(u_t, v) + a(t; u, v) = (f, v) \quad \forall t \in (0, T), \quad \forall v \in V, \quad v \text{ smooth.}$$

Taking for instance $v \in \mathcal{D}(\Omega)$ we obtain

$$\int_\Omega u_t v \, dx + \int_\Omega \left\{ a_{ij} \frac{\partial u}{\partial x_j} \frac{\partial v}{\partial x_i} + a_i \frac{\partial u}{\partial x_i} v + a_0 u v \right\} dx$$

$$= \int_\Omega f v \, dx \quad \forall t \in (0, T). \tag{8.34}$$

Integrating by parts in the second integral we get

$$\int_\Omega \left\{ u_t - \frac{\partial}{\partial x_i} \left(a_{ij} \frac{\partial u}{\partial x_j} \right) + a_i \frac{\partial u}{\partial x_i} + a_0 u \right\} v \, dx = \int_\Omega f \cdot v \, dx \quad \forall t \in (0, T),$$

i.e., it follows that

$$\frac{\partial u}{\partial t} - \frac{\partial}{\partial x_i}\left(a_{ij}\frac{\partial u}{\partial x_j}\right) + a_i\frac{\partial u}{\partial x_i} + a_0 u = f \text{ in } \Omega \times (0,T). \qquad (8.35)$$

Going back to (8.34) and taking a smooth $v \in V$ – but not necessarily vanishing on Γ – we get after the use of the Green formula for every $t \in (0,T)$:

$$\int_\Omega \left\{ u_t - \frac{\partial}{\partial x_i}\left(a_{ij}\frac{\partial u}{\partial x_j}\right) + a_i\frac{\partial u}{\partial x_i} + a_0 u\right\} v\, dx + \int_\Gamma a_{ij}\frac{\partial u}{\partial x_j}\nu_i v\, d\sigma(x) = \int_\Omega f \cdot v\, dx.$$

Using (8.35) we derive that

$$\int_\Gamma a_{ij}\frac{\partial u}{\partial x_j}\nu_i v\, d\sigma(x) = 0 \quad \forall v \in V \qquad (8.36)$$

v smooth. (Recall that $\nu = (\nu_1, \ldots, \nu_n)$ denotes the outward unit normal to Γ.) Of course to this we have to add the initial condition

$$u(x,0) = u_0(x). \qquad (8.37)$$

We can apply this to different cases:

- $V = H_0^1(\Omega)$.

In this case (8.33) is a weak formulation to the parabolic problem

$$\begin{cases} \dfrac{\partial u}{\partial t} - \dfrac{\partial}{\partial x_i}\left(a_{ij}\dfrac{\partial u}{\partial x_j}\right) + a_i\dfrac{\partial u}{\partial x_i} + a_0 u = f \text{ in } \Omega \times (0,T), \\ u(x,t) = 0 \quad \text{on } \Gamma \times (0,T), \\ u(x,0) = u_0(x) \quad \text{in } \Omega. \end{cases} \qquad (8.38)$$

When $a_{ij} = \delta_{ij}, \forall i,j = 1,\ldots,n, a_i = 0, \forall i = 0,\ldots,n$, we get the weak formulation of the classical heat equation

$$\begin{cases} \dfrac{\partial u}{\partial t} - \Delta u = f \quad \text{in } \Omega \times (0,T), \\ u(x,t) = 0 \quad \text{on } \Gamma \times (0,T), \\ u(x,0) = u_0(x) \quad \text{in } \Omega. \end{cases} \qquad (8.39)$$

- $V = H^1(\Omega)$.

Then – as seen in Chapter 1 – in this case (8.36) can be interpreted – for f suitable (see Chapter 1) as

$$a_{ij}\frac{\partial u}{\partial x_j}\nu_i = 0 \quad \text{on } \Gamma$$

and we have solved in a weak sense

$$
\begin{cases}
\dfrac{\partial u}{\partial t} - \dfrac{\partial}{\partial x_i}\left(a_{ij}\dfrac{\partial u}{\partial x_j}\right) + a_i\dfrac{\partial u}{\partial x_i} + a_0 u = f \text{ in } \Omega \times (0,T), \\[2mm]
a_{ij}(x)\dfrac{\partial u}{\partial x_j}\nu_i = 0 \quad \text{on } \Gamma \times (0,T), \\[2mm]
u(x,0) = u_0(x).
\end{cases}
\tag{8.40}
$$

This is a parabolic problem of Neumann type. In the particular case where $a_{ij} = \delta_{ij}$ $\forall\, i,j = 1,\dots,n$, $a_i = 0$ $\forall\, i = 0,\dots,n$ we have solved in a weak sense

$$
\begin{cases}
\dfrac{\partial u}{\partial t} - \Delta u = f \quad \text{in } \Omega \times (0,T), \\[2mm]
\dfrac{\partial u}{\partial n} = 0 \quad \text{on } \Gamma \times (0,T), \\[2mm]
u(x,0) = u_0(x) \quad \text{in } \Omega.
\end{cases}
\tag{8.41}
$$

Note that in the parabolic version of the Neumann problem $a_0 = 0$ is allowed.

• $V = \{\, v \in H^1(\Omega) \mid v = 0 \text{ on } \Gamma_0 \,\} = H_0^1(\Omega;\Gamma_0)$

where Γ_0 is some part of the boundary Γ. We denote by

$$
\Gamma_1 = \Gamma \setminus \Gamma_0
$$

the complementary part of Γ. Then, in this case, it is easy to see that (8.33) is a weak version of the problem

$$
\begin{cases}
\dfrac{\partial u}{\partial t} - \dfrac{\partial}{\partial x_i}\left(a_{ij}\dfrac{\partial u}{\partial x_j}\right) + a_i\dfrac{\partial u}{\partial x_i} + a_0 u = f \text{ in } \Omega \times (0,T), \\[2mm]
u = 0 \quad \text{on } \Gamma_0 \times (0,T), \\[2mm]
a_{ij}(x)\dfrac{\partial u}{\partial x_j}\nu_i = 0 \quad \text{on } \Gamma_1 \times (0,T), \\[2mm]
u(x,0) = u_0(x) \quad \text{in } \Omega.
\end{cases}
\tag{8.42}
$$

8.3. Nonlinear parabolic problems

Let Ω be a Lipschitz bounded open subset of $\mathbb{R}^n$ with boundary Γ. Denote by Γ_0 some measurable subset of Γ of positive measure (for the measure area $d\gamma(x)$) and by Γ_1 the complement of Γ_0 in Γ – that is to say

$$
\Gamma_1 = \Gamma \setminus \Gamma_0.
\tag{8.43}
$$

Set

$$
V = H_0^1(\Omega;\Gamma_0).
\tag{8.44}
$$

For $i,j = 1,\dots,n$ let $a_{ij}(x,t;u)$ be Carathéodory functions – i.e., such that

$$
u \mapsto a_{ij}(x,t;u) \text{ is continuous a.e. } (x,t) \in \Omega \times \mathbb{R}^+,
\tag{8.45}
$$

$$
(x,t) \mapsto a_{ij}(x,t;u), \text{ is measurable } \forall\, u \in \mathbb{R}.
\tag{8.46}
$$

(a_{ij} are defined on $\Omega \times \mathbb{R}^+ \times \mathbb{R}$, $\mathbb{R}^+ = (0, +\infty)$.) Moreover we assume that for some constants λ, Λ:

$$|a_{ij}(x, t; u)| \leq \Lambda \qquad \text{a.e. } (x, t) \in \Omega \times \mathbb{R}^+, \qquad \forall u \in \mathbb{R}, \tag{8.47}$$

$$a_{ij}(x, t; u)\xi_i\xi_j \geq \lambda|\xi|^2 \qquad \text{a.e. } (x, t) \in \Omega \times \mathbb{R}^+, \qquad \forall u \in \mathbb{R}, \forall \xi \in \mathbb{R}^n. \tag{8.48}$$

Then we would like to investigate the existence of a weak solution to the problem

$$\begin{cases} \dfrac{\partial u}{\partial t} - \dfrac{\partial}{\partial x_i}\left(a_{ij}(x, t; u)\dfrac{\partial u}{\partial x_j}\right) = f \text{ in } \Omega \times (0, T), \\[2mm] u = 0 \text{ on } \Gamma_0 \times (0, T), \\[2mm] u(x, 0) = u_0(x) \text{ in } \Omega, \end{cases} \tag{8.49}$$

where $u_0 \in L^2(\Omega)$, $f \in L^2(0, T; V')$. We have

THEOREM 8.10. *Under the above assumptions there exists a solution u to*

$$\begin{cases} u \in L^2(0, T; V), \ u_t \in L^2(0, T; V'), \\[2mm] u(0) = u_0, \\[2mm] \dfrac{d}{dt}(u, v) + a(u, v) = \langle f, v \rangle \text{ in } \mathcal{D}'(0, T), \ \forall v \in V. \end{cases} \tag{8.50}$$

$(\cdot, \cdot)$ *denotes the scalar product in $L^2(\Omega)$ and*

$$a(u, v) = \int_\Omega a_{ij}(x, t; u)\frac{\partial u}{\partial x_j} \cdot \frac{\partial v}{\partial x_i} \, dx \tag{8.51}$$

(clearly (8.50) is a weak formulation of (8.49)).

PROOF. We argue using the Schauder fixed point theorem. For that purpose, consider $w \in L^2(0, T; L^2(\Omega))$ and $u = T(w)$ the solution to

$$\begin{cases} u \in L^2(0, T; V), \ u_t \in L^2(0, T; V'), \\[2mm] u(0) = u_0, \\[2mm] \dfrac{d}{dt}(u, v) + a_w(u, v) = \langle f, v \rangle \text{ in } \mathcal{D}'(0, T), \ \forall v \in V, \end{cases} \tag{8.52}$$

where

$$a_w(u, v) = \int_\Omega a_{ij}(x, t; w)\frac{\partial u}{\partial x_j} \cdot \frac{\partial v}{\partial x_i} \, dx. \tag{8.53}$$

We know from Theorem 8.9 that such a $u = T(w)$ exists and is unique. If we can show that the mapping

$$w \mapsto T(w)$$

from $L^2(0, T; L^2(\Omega))$ into itself has a fixed point, we will be done.

First let us remark that by (8.27) for every $v \in L^2(0, T; V)$

$$\langle u_t, v \rangle + a_w(u, v) = \langle f, v \rangle \quad \text{a.e. } t \in (0, T). \tag{8.54}$$

Then taking $v = u$ in (8.54) we obtain

$$\frac{1}{2}\frac{d}{dt}(u,u) + a_w(u,u) = \langle f, u \rangle \quad \text{a.e. } t \in (0,T).$$

Using (8.48) it follows that

$$\frac{1}{2}\frac{d}{dt}(u,u) + \lambda \|u\|_V^2 \le \|f\|_* \|u\|_V \quad \text{a.e. } t \in (0,T) \qquad (8.55)$$

where we have set

$$\|u\|_V^2 = \int_\Omega |\nabla u|^2 \, dx, \qquad (8.56)$$

$$\|f\|_* = \underset{\|v\|_V \le 1}{\text{Sup}} |\langle f, v \rangle|. \qquad (8.57)$$

Using the Young inequality for the right-hand side of (8.55) – i.e.,

$$ab \le \frac{\lambda}{2}b^2 + \frac{1}{2\lambda}a^2,$$

we obtain

$$\frac{1}{2}\frac{d}{dt}(u,u) + \lambda \|u\|_V^2 \le \frac{\lambda}{2}\|u\|_V^2 + \frac{1}{2\lambda}\|f\|_*^2,$$

i.e.,

$$\frac{d}{dt}(u,u) + \lambda \|u\|_V^2 \le \frac{1}{\lambda}\|f\|_*^2 \quad \text{a.e. } t \in (0,T). \qquad (8.58)$$

Integrating on $(0,T)$ it follows that

$$|u(T)|_2^2 + \lambda \int_0^T \|u\|_V^2 \, dt \le |u_0|_2^2 + \frac{1}{\lambda}\int_0^T \|f\|_*^2 \, dt. \qquad (8.59)$$

In particular – due to (8.56) – we obtain easily

$$\|u\|_{L^2(0,T;V)}, \|u\|_{L^2(0,T;L^2(\Omega))} \le C \qquad (8.60)$$

with for some constant c

$$C^2 = c\{\frac{1}{\lambda}|u_0|_2^2 + \frac{1}{\lambda^2}\|f\|_{L^2(0,T;V')}^2\}. \qquad (8.61)$$

Setting

$$B = \{ v \in L^2(0,T;L^2(\Omega)) \mid \|v\|_{L^2(0,T;L^2(\Omega))} \le C \} \qquad (8.62)$$

it is clear from (8.60) that

$$w \mapsto u = T(w)$$

is a mapping from B into itself. Moreover, going back to the third equation of (8.52) – see also (8.54) – we have

$$\|u_t\|_{L^2(0,T;V')} \le M\|u\|_{L^2(0,T;V)} + \|f\|_{L^2(0,T;V')}$$
$$\le MC + \|f\|_{L^2(0,T;V')} = C'. \qquad (8.63)$$

Thus, u belongs to a bounded subset of $H^1(0,T;V,V')$ which is relatively compact in $L^2(0,T;L^2(\Omega))$ (see [17]). Thus, in order to conclude by applying the Schauder fixed point theorem in B, we just need to show that T is continuous on B. So, let us consider a sequence $w_n \in B$ such that

$$w_n \to w \quad \text{in } B.$$

Let us denote by $u_n = T(w_n)$ the solution to (8.52) corresponding to $w = w_n$. Due to (8.60) and (8.63) we can extract from n a subsequence, still labeled by n, such that

$$
\begin{aligned}
w_n &\to w & \text{a.e. in } \Omega \times (0,T), \\
u_n &\to u_\infty & \text{in } L^2(0,T;L^2(\Omega)), \\
\frac{\partial u_n}{\partial x_i} &\rightharpoonup \frac{\partial u_\infty}{\partial x_i} & \text{in } L^2(0,T;L^2(\Omega)), \\
u_{n_t} &\rightharpoonup u_{\infty_t} & \text{in } L^2(0,T;V'),
\end{aligned}
\tag{8.64}
$$

where $u_\infty \in L^2(0,T;V)$. From the third equation of (8.52) we deduce

$$
\int_0^T -(u_n,v)\varphi'(t)\,dt + \int_0^T \int_\Omega \varphi(t) \cdot a_{ij}(x,t;w_n)\frac{\partial u_n}{\partial x_j} \cdot \frac{\partial v}{\partial x_i}\,dx\,dt
$$
$$
= \int_0^T \langle f, v\rangle \varphi(t)\,dt \quad \forall\, v \in V, \; \forall\, \varphi \in \mathcal{D}(0,T).
\tag{8.65}
$$

By the Lebesgue convergence theorem we have

$$
\varphi(t)a_{ij}(x,t;w_n)\frac{\partial v}{\partial x_j} \to \varphi(t)a_{ij}(x,t;w)\frac{\partial v}{\partial x_j},
$$

in $L^2(0,T;L^2(\Omega)) = L^2(\Omega \times (0,T))$. Thus passing to the limit in (8.65) leads to

$$
\frac{d}{dt}(u_\infty,v) + a_w(u_\infty,v) = \langle f, v\rangle \quad \text{in } \mathcal{D}'(0,T).
\tag{8.66}
$$

Of course

$$
u_\infty \in L^2(0,T;V), \qquad u_{\infty,t} \in L^2(0,T;V').
\tag{8.67}
$$

Now, we can write for a.e. $t \in (0,T)$, $\forall\, v \in V$

$$
(u_n(t),v) - (u_0,v) = \int_0^t \langle u_{n_t}, v\rangle\,dt.
$$

(Recall that by (8.60), (8.63), (8.64), $u_n, u_\infty \in C([0,T],L^2(\Omega))$ – see also (8.17) – and if we pass to the limit we get

$$
(u_\infty(t),v) - (u_0,v) = \int_0^t \langle u_{\infty_t}, v\rangle\,dt = (u_\infty(t),v) - (u_\infty(0),v)
$$

a.e. $t \in (0,T)$, $\forall\, v \in V$. Note that by the second line of (8.64) one can assume without loss of generality that

$$
u_n(t) \to u_\infty(t) \quad \text{in } L^2(\Omega), \text{ a.e. } t \in (0,T).
$$

Thus

$$
u_\infty(0) = u_0.
\tag{8.68}
$$

Combining (8.66)–(8.68) one sees that $u_\infty = T(w)$ – since u_n has only $T(w)$ as possible limit,

$$u_n = T(w_n) \to T(w)$$

in $L^2(0, T; L^2(\Omega))$ and the continuity of T is established.   □

REMARK 8.3. It is possible to assume that $|\Gamma_0| = 0$. Indeed, if we set

$$\tilde{u} = e^{-kt}u,$$

then u is a solution to (8.49) iff $\tilde{u}$ satisfies.

$$\begin{cases} \dfrac{\partial}{\partial t}(e^{kt}\tilde{u}) - \dfrac{\partial}{\partial x_i}\left(a_{ij}(x, t; e^{kt}\tilde{u})e^{kt}\dfrac{\partial \tilde{u}}{\partial x_j}\right) = f, \\[2mm] \tilde{u} = 0 \text{ on } \Gamma_0 \times (0, T), \\[2mm] \tilde{u}(0) = u_0. \end{cases}$$

The first above equation can also be written

$$\frac{\partial \tilde{u}}{\partial t} + k\tilde{u} - \frac{\partial}{\partial x_i}\left(a_{ij}(x, t; e^{kt}\tilde{u})\frac{\partial \tilde{u}}{\partial x_j}\right) = fe^{-kt}.$$

Thus, choosing k large enough due to (8.48), we can always assume that (8.55) holds with $\| \ \|_V$ being the H^1 norm.

We now turn to the question of uniqueness. For this purpose we consider the problem (8.50). Regarding a_{ij}, we apply the assumptions of Theorem 8.10. In addition we will suppose that

$$|a_{ij}(x, t; u) - a_{ij}(x, t; \hat{u})| \leq C\omega(|u - \hat{u}|) \tag{8.69}$$

for every $u, \hat{u} \in \mathbb{R}$, where C is a constant and ω is a continuous positive function satisfying

$$\int_{0+} \frac{ds}{\omega^2(s)} = +\infty. \tag{8.70}$$

(ω can be the supremum of the moduli of continuity of the a_{ij}'s.)

Then under the above assumptions we can show

THEOREM 8.11. _If the above assumptions hold, then there exists a unique solution to (8.50)._

PROOF. Let $u, \hat{u}$ be two solutions to (8.50). We know that, in a weak sense in $L^2(0, T; V')$,

$$\frac{\partial u}{\partial t} - \frac{\partial}{\partial x_i}\left(a_{ij}(x, t; u)\frac{\partial u}{\partial x_j}\right) = \frac{\partial \hat{u}}{\partial t} - \frac{\partial}{\partial x_i}\left(a_{ij}(x, t; \hat{u})\frac{\partial \hat{u}}{\partial x_j}\right)$$

in $\Omega \times (0, T)$. This can also be written

$$\frac{\partial}{\partial t}(u-\hat{u}) - \frac{\partial}{\partial x_i}\left(a_{ij}(x, t; u)\frac{\partial}{\partial x_j}(u-\hat{u})\right) = -\frac{\partial}{\partial x_i}\left([a_{ij}(x, t; \hat{u}) - a_{ij}(x, t; u)]\frac{\partial \hat{u}}{\partial x_j}\right). \tag{8.71}$$

Without loss of generality we can assume that

$$0 < \omega(s) \quad \forall s > 0, \qquad \int_1^{+\infty} \frac{ds}{\omega^2(s)} < +\infty. \tag{8.72}$$

Then, for $\varepsilon > 0$, we set

$$H_\varepsilon(\xi) = \begin{cases} 0 & \text{if } \xi \le \varepsilon, \\ \dfrac{1}{I_\varepsilon} \displaystyle\int_\varepsilon^\xi \frac{ds}{\omega^2(s)} & \text{if } \xi \ge \varepsilon, \end{cases} \tag{8.73}$$

where

$$I_\varepsilon = \int_\varepsilon^{+\infty} \frac{ds}{\omega^2(s)}. \tag{8.74}$$

For $\xi > 0$ and ε small enough we have $\varepsilon < \xi$ and

$$H_\varepsilon(\xi) = 1 - \frac{1}{I_\varepsilon} \int_\xi^{+\infty} \frac{ds}{\omega^2(s)} \to 1, \tag{8.75}$$

as $\varepsilon \to 0$ (see (8.70)). Moreover

$$H_\varepsilon(\xi) = 0 \tag{8.76}$$

for $\xi \le 0$. We consider then the function

$$H_\varepsilon(u - \hat{u})$$

that belongs to V. Using this function with (8.71), we get easily

$$\langle (u - \hat{u})_t, H_\varepsilon(u - \hat{u}) \rangle + \int_\Omega \lambda |\nabla(u - \hat{u})|^2 H_\varepsilon'(u - \hat{u}) \, dx$$
$$\le \int_\Omega (A(\cdot; \hat{u}) - A(\cdot; u)) \nabla \hat{u} \cdot \nabla(u - \hat{u}) H_\varepsilon'(u - \hat{u}) \, dx,$$

where A denotes the matrix with entries a_{ij}. Using now (8.69) we get for some constant C:

$$\langle (u - \hat{u})_t, H_\varepsilon(u - \hat{u}) \rangle + \lambda \int_\Omega |\nabla(u - \hat{u})|^2 H_\varepsilon' \, dx$$
$$\le C \int_\Omega \omega(u - \hat{u}) |\nabla \hat{u}| |\nabla(u - \hat{u})| H_\varepsilon'(u - \hat{u}) \, dx.$$

Note that by the definition of H_ε we are integrating on the set

$$S = [u - \hat{u} > \varepsilon] = \{ x \in \Omega \mid (u - \hat{u})(x) > \varepsilon \}.$$

Thus it follows that

$$\langle (u - \hat{u})_t, H_\varepsilon(u - \hat{u}) \rangle + \frac{\lambda}{I_\varepsilon} \int_S \frac{|\nabla(u - \hat{u})|^2}{\omega^2(u - \hat{u})} \, dx \le \frac{C}{I_\varepsilon} \int_S |\nabla \hat{u}| \cdot \frac{|\nabla(u - \hat{u})|}{\omega(u - \hat{u})} \, dx.$$

Using now the Young inequality

$$ab \le \frac{\lambda}{2C} a^2 + \frac{C}{2\lambda} b^2 \quad \forall a, b > 0$$

in the last integral it follows easily that

$$\langle (u - \hat{u})_t, H_\varepsilon(u - \hat{u}) \rangle + \frac{\lambda}{2I_\varepsilon} \int_S \frac{|\nabla(u - \hat{u})|^2}{\omega^2(u - \hat{u})} \, dx \leq \frac{C^2}{2\lambda I_\varepsilon} \int_S |\nabla \hat{u}|^2 \, dx$$

from which follows

$$\langle (u - \hat{u})_t, H_\varepsilon(u - \hat{u}) \rangle \leq \frac{C^2}{2\lambda I_\varepsilon} \int_\Omega |\nabla \hat{u}|^2 \, dx. \tag{8.77}$$

Setting

$$K_\varepsilon(\xi) = \int_0^\xi H_\varepsilon(s) \, ds,$$

this inequality can also be written

$$\frac{d}{dt} \int_\Omega K_\varepsilon(u - \hat{u}) \, dx \leq \frac{C^2}{2\lambda I_\varepsilon} \int_\Omega |\nabla \hat{u}|^2 \, dx.$$

Integrating on $(0, t)$ we get for every t

$$\int_\Omega K_\varepsilon(u - \hat{u}) \, dx \leq \frac{C^2}{2\lambda I_\varepsilon} \int_0^t \int_\Omega |\nabla \hat{u}|^2 \, dx. \tag{8.78}$$

Now due to (8.75), (8.76), when $\varepsilon \to 0$

$$K_\varepsilon(\xi) \to \xi \vee 0, \qquad I_\varepsilon \to +\infty$$

where $\vee$ denotes the maximum of numbers.

Letting $\varepsilon \to 0$ in (8.78) leads to

$$\int_\Omega [u - \hat{u}]^+(x, t) \, dx \leq 0$$

for a.e. t. Thus we obtain

$$u - \hat{u} \leq 0.$$

Exchanging the roles of u and $\hat{u}$ leads to the result. □

REMARK 8.4. If $u_0, \hat{u}_0 \in L^2(\Omega)$ and if $u, \hat{u}$ denote the solutions to (8.50) corresponding to u_0 and $\hat{u}_0$ respectively, the above proof shows in fact that

$$\int_\Omega [u - \hat{u}]^+(x, t) \, dx \leq \int_\Omega [u_0 - \hat{u}_0]^+ \, dx.$$

In particular $u_0 \leq \hat{u}_0$ would imply

$$u \leq \hat{u}, \tag{8.79}$$

i.e. the solution to (8.50) is monotone in u_0. Note also that using the same arguments as above we can show that the solution is monotone in f as well.

Chapter 9

Asymptotic Behaviour of Parabolic Problems

We first consider the linear case following some of the lines of [**12**].

9.1. The linear case

As in the previous chapters, we consider a bounded open subset of $\mathbb{R}^n$ given by

$$\Omega_\ell = (-\ell, \ell)^p \times \omega \tag{9.1}$$

where ω is a bounded open subset of $\mathbb{R}^{n-p}$. For $x = (x_1, \ldots, x_n)$ we set as before

$$\mathrm{X}_1 = (x_1, \ldots, x_p), \qquad \mathrm{X}_2 = (x_{p+1}, \ldots, x_n). \tag{9.2}$$

If $\partial_0 \omega$ denotes a measurable subset of $\partial \omega$ – the boundary of ω – we introduce $H_0^1(\omega; \partial_0\omega)$ the space defined in Chapter 1 (see (1.28), (1.36)). Note that in this section $\partial_0\omega$ can be of superficial measure 0. Then, we set

$$\Gamma_0 = \partial(-\ell, \ell)^p \times \omega \cup (-\ell, \ell)^p \times \partial_0\omega \tag{9.3}$$

and introduce the space $H_0^1(\Omega_\ell; \Gamma_0)$. If $H_0^1(\omega; \partial_0\omega)'$ denotes the dual space of $H_0^1(\omega; \partial_0\omega)$ it is possible to extend

$$f \in L^2(0, T; H_0^1(\omega; \partial_0\omega)') \tag{9.4}$$

into a distribution of $L^2(0, T; H_0^1(\Omega_\ell; \Gamma_0)')$. Indeed, for any $v \in H_0^1(\Omega_\ell; \Gamma_0)$ we set

$$\langle \tilde{f}(t), v \rangle = \int_{(-\ell,\ell)^p} \langle f(t), v(\mathrm{X}_1, \cdot) \rangle \, d\mathrm{X}_1. \tag{9.5}$$

In the above integral the duality bracket is the duality between $H_0^1(\omega; \partial_0\omega)'$ and $H_0^1(\omega; \partial_0\omega)$, recall that for $v \in H_0^1(\Omega_\ell; \Gamma_0)$, $v(\mathrm{X}_1, \cdot) \in H_0^1(\omega, \partial_0\omega)$ (by Proposition 3.1). The duality in the left-hand side of (9.5) is the duality between $H_0^1(\Omega_\ell; \Gamma_0)'$, $H_0^1(\Omega_\ell; \Gamma_0)$ as we will see below. We have indeed:

PROPOSITION 9.1. *The formula (9.5) makes sense for almost every $t \in (0, T)$ and defines an element $\tilde{f}(t) \in H_0^1(\Omega_\ell; \Gamma_0)'$. Moreover it holds that*

$$\tilde{f} \in L^2(0, T; H_0^1(\Omega_\ell; \Gamma_0)'). \tag{9.6}$$

REMARK 9.1. Before giving the proof of Proposition 9.1, let us notice that when f and v are smooth functions it holds that

$$\langle \tilde{f}(t), v \rangle = \int_{\Omega_\ell} f(t, \mathrm{X}_2) v(x) \, dx$$

i.e., $\tilde{f}$ is the natural extension of f as a function.

PROOF OF PROPOSITION 9.1. Let us first show that (9.5) is well defined for a.e. $t \in (0,T)$. For $v \in H_0^1(\Omega_\ell; \Gamma_0)$ it holds that for a.e. $t \in (0,T)$

$$|\langle f(t), v(\mathbf{x}_1, \cdot)\rangle| \leq |f(t)|_{*,\omega} \|v(\mathbf{x}_1, \cdot)\|_{1,2,\omega} \tag{9.7}$$

where $|\cdot|_{*,\omega}$ denotes the strong dual norm in $H_0^1(\omega; \partial_0\omega)'$. $\|\cdot\|_{1,2,\omega}$ is the usual $H^1(\omega)$ norm. We claim that the function

$$\mathbf{x}_1 \mapsto \langle f(t), v(\mathbf{x}_1, \cdot)\rangle \tag{9.8}$$

is measurable. Indeed, consider a sequence of functions $v_n \in C_0^1(\overline{\Omega_\ell}; \Gamma_0)$ such that

$$v_n \to v \quad \text{in } H_0^1(\Omega_\ell; \Gamma_0). \tag{9.9}$$

Then, see Proposition 3.1, for almost every $\mathbf{x}_1 \in (-\ell, \ell)^p$ it holds that

$$v_n(\mathbf{x}_1, \cdot) \to v(\mathbf{x}_1, \cdot) \quad \text{in } H_0^1(\omega; \partial_0\omega). \tag{9.10}$$

It follows from (9.7) that

$$\langle f(t), v_n(\mathbf{x}_1, \cdot)\rangle \to \langle f(t), v(\mathbf{x}_1, \cdot)\rangle \tag{9.11}$$

for a.e. $\mathbf{x}_1 \in (-\ell, \ell)^p$ and a.e. $t \in (0,T)$. Thus, the function defined by (9.8) will be measurable if for every n the function

$$\mathbf{x}_1 \mapsto \langle f(t), v_n(\mathbf{x}_1, \cdot)\rangle$$

is. But, it follows easily from (9.7), that since $v_n \in C_0^1(\overline{\Omega_\ell}; \Gamma_0)$ this function is continuous and thus measurable. Going back to (9.7) we see that

$$|\langle f(t), v(\mathbf{x}_1, \cdot)\rangle|^2 \leq |f|_{*,\omega}^2 \|v(\mathbf{x}_1, \cdot)\|_{1,2,\omega}^2 \in L^1((-\ell, \ell)^p). \tag{9.12}$$

Thus – since $(-\ell, \ell)^p$ is a bounded domain – it follows that the function defined by (9.8) belongs to $L^1((-\ell, \ell)^p)$ and for almost every $t \in (0,T)$ the integral in (9.5) makes sense. Moreover, from (9.7) we derive

$$
\begin{aligned}
|\langle \tilde{f}(t), v\rangle| &= \left| \int_{(-\ell,\ell)^p} \langle f(t), v(\mathbf{x}_1, \cdot)\rangle \, d\mathbf{x}_1 \right| \\
&\leq \int_{(-\ell,\ell)^p} |\langle f(t), v(\mathbf{x}_1, \cdot)\rangle| \, d\mathbf{x}_1 \\
&\leq |f(t)|_{*,\omega} \int_{(-\ell,\ell)^p} \|v(\mathbf{x}_1, \cdot)\|_{1,2,\omega} \, d\mathbf{x}_1 \\
&\leq |f(t)|_{*,\omega} \left(\int_{(-\ell,\ell)^p} \|v(\mathbf{x}_1, \cdot)\|_{1,2,\omega}^2 \, d\mathbf{x}_1 \right)^{1/2} (2\ell)^{p/2} \\
&\leq |f(t)|_{*,\omega} (2\ell)^{p/2} \|v\|_{1,2,\Omega_\ell}.
\end{aligned}
\tag{9.13}
$$

This shows that $\tilde{f}(t)$ defined by (9.5) belongs to $H_0^1(\Omega_\ell; \Gamma_0)'$. Moreover, from (9.13) it holds that

$$|\tilde{f}(t)|_{*,\Omega_\ell} \leq |f(t)|_{*,\omega} (2\ell)^{p/2} \tag{9.14}$$

if we use the same notation to denote the strong dual norms in $H_0^1(\omega; \partial_0\omega)'$ and $H_0^1(\Omega_\ell; \Gamma_0)'$. If $f \in L^2(0, T; H_0^1(\omega, \partial_0\omega)')$, since f is measurable, there exists a sequence of simple functions $f_n \in L^2(0, T; H_0^1(\omega, \partial_0\omega)')$ such that, a.e. $t \in (0, T)$,

$$f_n(t) \to f(t) \quad \text{in } H_0^1(\omega, \partial_0\omega)'. \tag{9.15}$$

It follows then that $\tilde{f}_n(t)$ defined by (9.5) with f_n in place of f is a sequence of simple functions with values in $H_0^1(\Omega_\ell; \Gamma_0)'$ such that

$$\tilde{f}_n(t) \to \tilde{f}(t) \quad \text{a.e. } t \in (0, T) \text{ in } H_0^1(\Omega_\ell; \Gamma_0)' \tag{9.16}$$

(see (9.14)). Thus $\tilde{f}$ is a measurable function from $(0, T)$ into $H_0^1(\Omega_\ell; \Gamma_0)'$. Moreover, from (9.14) we derive

$$|\tilde{f}(t)|_{*,\Omega_\ell}^2 \le |f(t)|_{*,\omega}^2 (2\ell)^p. \tag{9.17}$$

It follows that $\tilde{f}(t) \in L^2(0, T; H_0^1(\Omega_\ell; \Gamma_0)')$ with

$$\int_0^T |\tilde{f}(t)|_{*,\Omega_\ell}^2 \, dt \le (2\ell)^p \int_0^T |f(t)|_{*,\omega}^2 \, dt \tag{9.18}$$

i.e., with an obvious notation for the norm introduced

$$|\tilde{f}|_{L^2(0,T;H_0^1(\Omega_\ell;\Gamma_0)')} \le (2\ell)^{p/2} |f|_{L^2(0,T;H_0^1(\omega;\partial_0\omega)')}. \tag{9.19}$$

This completes the proof of Proposition 9.1. $\qquad\qquad\qquad\qquad\qquad\square$

REMARK 9.2. In the following, when no confusion occurs, we will denote $\tilde{f}$ simply by f as it is done in the case where f is a function (see Remark 9.1).

We introduce now coefficients a_{ij} such that

$$a_{ij} = a_{ij}(x, t) \in L^\infty(\mathbb{R}^p \times \omega \times (0, T)) \quad \forall i, j = 1, \ldots, n. \tag{9.20}$$

To simplify our notation we set

$$A = A(x, t) = (a_{ij})_{i,j=1,\ldots,n} \tag{9.21}$$

and we suppose that

$$|A(x, t)| \le \Lambda \qquad\qquad \text{a.e. } x \in \mathbb{R}^p \times \omega, \qquad \text{a.e. } t \in (0, T), \tag{9.22}$$

$$A(x, t)\xi\xi \ge \lambda|\xi|^2 \quad \forall \xi \in \mathbb{R}^n, \qquad \text{a.e. } x \in \mathbb{R}^p \times \omega, \qquad \text{a.e. } t \in (0, T). \tag{9.23}$$

(In the above formulae $|\xi|$ denotes the euclidean norm of $\xi \in \mathbb{R}^n$, $|A|$ the norm of the matrix A subordinated to the euclidean norm.) Using the splitting of A in form of submatrices at the level of the p-th column and the p-th line (see also (3.24)) we assume in addition that

$$A = \begin{pmatrix} A_{11}(x, t) & A_{12}(\mathrm{x}_2, t) \\ A_{21}(x, t) & A_{22}(\mathrm{x}_2, t) \end{pmatrix} \tag{9.24}$$

that is to say we suppose that the $n - p$ last columns of A depend on x_2, t only.

Let us recall (see Chapter 8) that

$$H^1(0, T; V, V') = \{ v \in L^2(0, T; V) \mid v_t \in L^2(0, T; V') \}. \tag{9.25}$$

Then, we have seen in the preceding chapter, that for $f \in L^2(0, T; H_0^1(\omega, \partial_0\omega)')$, $u_0 \in L^2(\omega)$ there exists a unique u_∞ solution to

$$\begin{cases} u_\infty \in H^1(0, T; H_0^1(\omega; \partial_0\omega), H_0^1(\omega; \partial_0\omega)'), \\ \dfrac{d}{dt}(u_\infty, v)_{2,\omega} + \displaystyle\int_\omega A_{22}\nabla_{X_2} u_\infty \nabla_{X_2} v \, dX_2 = \langle f, v \rangle \quad \text{in } \mathcal{D}'(0, T), \\ \qquad\qquad \forall\, v \in H_0^1(\omega; \partial_0\omega), \\ u_\infty(\cdot, 0) = u_0 \quad \text{in } \omega. \end{cases} \tag{9.26}$$

$(\cdot, \cdot)_{2,K}$ denotes the usual scalar product in $L^2(K)$, K any measurable subset of $\mathbb{R}^n$. Moreover, if $\tilde{f} = f$ denotes the extension of f defined in Proposition 9.1 there exists a unique u_ℓ solution to

$$\begin{cases} u_\ell \in H^1(0, T; H_0^1(\Omega_\ell; \Gamma_0), H_0^1(\Omega_\ell; \Gamma_0)'), \\ \dfrac{d}{dt}(u_\ell, v)_{2,\Omega_\ell} + \displaystyle\int_{\Omega_\ell} A\nabla u_\ell \nabla v \, dx = \langle f, v \rangle \quad \text{in } \mathcal{D}'(0, T), \\ \qquad\qquad \forall\, v \in H_0^1(\Omega_\ell; \Gamma_0), \\ u_\ell(\cdot, 0) = u_0 \quad \text{in } \Omega_\ell. \end{cases} \tag{9.27}$$

Then, as expected, we would like to show that $u_\ell \to u_\infty$. The spaces where this convergence occurs will be made precise below. Note that we did not consider here lower order terms – i.e., with the notation of Chapter 3, we took $a = 0$ only for the sake of simplicity and such lower order terms will not change anything here. Thus, let us establish

THEOREM 9.2. *For any $\ell_0 > 0$, any $r > 0$ there exists a constant C independent of ℓ such that*

$$|u_\ell - u_\infty|_{L^\infty(0,T;L^2(\Omega_{\ell_0}))} + |u_\ell - u_\infty|_{L^2(0,T;H^1(\Omega_{\ell_0}))} \leq \frac{C}{\ell^r}. \tag{9.28}$$

REMARK 9.3. As we mentioned above, for the sake of simplicity we consider here only the case where $a = 0$. Rescaling the time – see Remark 8.3 of the preceding chapter – we can always assume the existence of a lower order term in au with $a \geq \lambda > 0$. However, to simplify our exposition we will suppose below that

$$|\partial_0\omega| > 0 \tag{9.29}$$

i.e., the superficial measure of the above set is positive. The reader will easily establish the results in the case where (9.29) does not hold.

Before giving the proof of Theorem 9.2 let us establish the following lemma:

LEMMA 9.3 (Estimate of u_ℓ). *Let us denote by u_ℓ the solution to (9.27). There exists a constant $C = C(\lambda, p)$ such that*

$$\begin{aligned} |u_\ell|_{L^\infty(0,T;L^2(\Omega_\ell))}^2 &+ \int_0^T \|\nabla u_\ell\|_{2,\Omega_\ell}^2 \, dt \\ &\leq C\ell^p \{|u_0|_{2,\omega}^2 + |f|_{L^2(0,T;H_0^1(\omega,\partial_0\omega)')}^2\}. \end{aligned} \tag{9.30}$$

PROOF. We take $v = u_\ell$ in (9.27). We obtain for a.e. t

$$\frac{d}{dt}\frac{1}{2}|u_\ell|_{2,\Omega_\ell}^2 + \int_{\Omega_\ell} A\nabla u_\ell \nabla u_\ell \, dx = \langle f, u_\ell \rangle \quad \text{a.e. } t \in (0,T). \tag{9.31}$$

Using (9.14), (9.23) we obtain

$$\frac{1}{2}\frac{d}{dt}|u_\ell|_{2,\Omega_\ell}^2 + \lambda \int_{\Omega_\ell} |\nabla u_\ell|^2 \, dx \leq |f|_{*,\Omega_\ell}||\nabla u_\ell||_{2,\Omega_\ell}. \tag{9.32}$$

REMARK 9.4. In the above formula $|\cdot|_{*,\Omega_\ell}$ denotes the strong dual norm of $H_0^1(\Omega_\ell;\Gamma_0)'$ when one equips $H_0^1(\Omega_\ell;\Gamma_0)$ with the norm $||\nabla u_\ell||_{2,\Omega_\ell}$. It is easy to see that if $|\cdot|_*$ denotes the strong dual norm of $H_0^1(\omega;\partial_0\omega)'$ when $H_0^1(\omega;\partial_0\omega)$ is equipped with the norm $||\nabla_{X_2} v||_{2,\omega}$, then it holds that

$$|f|_{*,\Omega_\ell} \leq |f|_*(2\ell)^{p/2} \tag{9.33}$$

(simply reproduce the arguments of Proposition 9.1).

Using the Young inequality in (9.32) we obtain – see (9.33) –

$$\frac{1}{2}\frac{d}{dt}|u_\ell|_{2,\Omega_\ell}^2 + \lambda||\nabla u_\ell||_{2,\Omega_\ell}^2 \leq \frac{\lambda}{2}||\nabla u_\ell||_{2,\Omega_\ell}^2 + \frac{1}{2\lambda}|f|_{*,\Omega_\ell}^2$$
$$\leq \frac{\lambda}{2}||\nabla u_\ell||_{2,\Omega_\ell}^2 + \frac{1}{2\lambda}|f|_*^2(2\ell)^p. \tag{9.34}$$

Simplifying by $\frac{1}{2}$ and integrating on $(0,t)$ we obtain

$$|u_\ell|_{2,\Omega_\ell}^2(t) + \lambda \int_0^t ||\nabla u_\ell||_{2,\Omega_\ell}^2 \, dt \leq |u_0|_{2,\Omega_\ell}^2 + \frac{(2\ell)^p}{\lambda}\int_0^t |f|_*^2 \, dt. \tag{9.35}$$

Clearly

$$|u_0|_{2,\Omega_\ell}^2 = \int_{(-\ell,\ell)^p}\int_\omega u_0(X_1, X_2)^2 \, dX_2 \, dX_1 = (2\ell)^p|u_0|_{2,\omega}^2. \tag{9.36}$$

Thus, from (9.35) we derive for a.e. $t \in (0,T)$:

$$|u_\ell|_{2,\Omega_\ell}^2(t) + \lambda \int_0^t ||\nabla u_\ell||_{2,\Omega_\ell}^2 \, dt \leq (2\ell)^p\left\{|u_0|_{2,\omega}^2 + \frac{1}{\lambda}\int_0^T |f|_*^2 \, dt\right\}$$

which implies

$$\min(1,\lambda)\left\{|u_\ell|_{2,\Omega_\ell}^2(t) + \int_0^T ||\nabla u_\ell||_{2,\Omega_\ell}^2 \, dt\right\}$$
$$\leq (2\ell)^p\left\{|u_0|_{2,\omega}^2 + \frac{1}{\lambda}|f|_{L^2(0,T;H_0^1(\omega,\partial_0\omega)')}^2\right\}.$$

The inequality (9.30) follows then easily. □

Then we can turn to the proof of Theorem 9.2.

PROOF OF THEOREM 9.2. As usual we consider ϱ a smooth function of $\mathbb{R}^p$ such that

$$0 \leq \varrho \leq 1, \qquad \varrho = 1 \quad \text{on} \quad \left(-\frac{1}{2}, \frac{1}{2}\right)^p, \qquad \varrho = 0 \quad \text{outside} \quad (-1,1)^p, \qquad (9.37)$$

$$|\nabla_{X_1} \varrho| \leq \theta \tag{9.38}$$

where θ is some positive constant. Then, for a.e. $t \in (0,T)$, $\ell_1 \leq \ell$ it holds that

$$(u_\ell - u_\infty)\varrho^2\left(\frac{X_1}{\ell_1}\right) \in H_0^1(\Omega_\ell; \Gamma_0). \tag{9.39}$$

From (9.26), (9.27) we then derive (see (9.5))

$$\left\langle \frac{du_\ell}{dt}, (u_\ell - u_\infty)\varrho^2 \right\rangle + \int_{\Omega_\ell} A\nabla u_\ell \nabla\{(u_\ell - u_\infty)\varrho^2\}\, dx$$

$$= \langle f, (u_\ell - u_\infty)\varrho^2 \rangle$$

$$= \int_{(-\ell,\ell)^p} \left\{ \left\langle \frac{du_\infty}{dt}, (u_\ell - u_\infty)\varrho^2 \right\rangle \right. \tag{9.40}$$

$$\left. + \int_\omega A_{22}\nabla_{X_2} u_\infty \nabla_{X_2}(u_\ell - u_\infty)\varrho^2\, dX_2 \right\} dX_1.$$

(For simplicity we have denoted $\varrho^2(\frac{X_1}{\ell_1})$ by ϱ^2.) We need then some technical results. Namely first we claim that

- $\varrho u_\ell \in H^1(0,T;V,V')$ where we have set $V = H_0^1(\Omega_\ell; \Gamma_0)$ and it holds that

$$\left\langle \frac{d}{dt}(\varrho u_\ell), v \right\rangle = \left\langle \frac{du_\ell}{dt}, \varrho v \right\rangle \quad \forall v \in H_0^1(\Omega_\ell; \Gamma_0), \text{ in } \mathcal{D}'(0,T). \tag{9.41}$$

Indeed, due to the definition of ϱ we have clearly

$$\varrho u_\ell \in L^2(0,T; H_0^1(\Omega_\ell; \Gamma_0)). \tag{9.42}$$

Moreover, for any $\varphi \in \mathcal{D}(0,T)$, $v \in H_0^1(\Omega_\ell; \Gamma_0)$ it holds that

$$-\int_{\Omega_\ell} v \int_0^T \varrho u_\ell \varphi'\, dt\, dx = -\int_{\Omega_\ell} \varrho v \int_0^T u_\ell \varphi'\, dt\, dx$$

$$= \int_0^T \left\langle \frac{du_\ell}{dt}, \varrho v \right\rangle \varphi\, dt. \tag{9.43}$$

Clearly, the mapping

$$v \mapsto \left\langle \frac{du_\ell}{dt}, \varrho v \right\rangle$$

is linear continuous on $H_0^1(\Omega_\ell; \Gamma_0)$ and we have

$$\left| \left\langle \frac{du_\ell}{dt}, \varrho v \right\rangle \right| \leq \left| \frac{du_\ell}{dt} \right|_{V'} \|\nabla \varrho v\|_{2,\Omega_\ell} \leq C(\varrho) \left| \frac{du_\ell}{dt} \right|_{V'} \|\nabla v\|_{2,\Omega_\ell} \qquad (9.44)$$

where $|\cdot|_{V'}$ denotes the strong dual norm on V', for instance subordinated to the norm $\|\nabla v\|_{2,\Omega_\ell}$ of V. $C(\varrho)$ is some constant depending on ϱ only. The claim (9.41) follows then easily.

We have also, recalling (9.5)

- $\varrho u_\infty \in H^1(0,T;V,V')$ and it holds that (in $\mathcal{D}'(0,T)$)

$$\left\langle \frac{d}{dt}(\varrho u_\infty), v \right\rangle = \left\langle \widetilde{\frac{du_\infty}{dt}}, \varrho v \right\rangle \quad \forall v \in H_0^1(\Omega_\ell; \Gamma_0). \qquad (9.45)$$

It is clear that

$$\varrho u_\infty \in L^2(0,T;H_0^1(\Omega_\ell; \Gamma_0)). \qquad (9.46)$$

Moreover, for any $\varphi \in \mathcal{D}(0,T)$ and $v \in H_0^1(\Omega_\ell; \Gamma_0)$ it holds that

$$
\begin{aligned}
-\int_{\Omega_\ell} v \int_0^T \varrho u_\infty \varphi' \, dt \, dx &= -\int_0^T \int_{(-\ell,\ell)^p} \int_\omega u_\infty \varrho v \varphi' \, dx_2 \, dx_1 \, dt \\
&= \int_0^T \int_{(-\ell,\ell)^p} \left\langle \frac{du_\infty}{dt}, \varrho v(x_1, \cdot) \right\rangle \varphi \, dt \, dx_1 \qquad (9.47) \\
&= \int_0^T \left\langle \widetilde{\frac{du_\infty}{dt}}, \varrho v \right\rangle \varphi \, dt.
\end{aligned}
$$

We can then easily conclude as above.

Thanks to these technical results we derive that

$$
\begin{aligned}
\int_0^t \left\{ \left\langle \frac{du_\ell}{dt}, (u_\ell - u_\infty)\varrho^2 \right\rangle - \int_{(-\ell,\ell)^p} \left\langle \frac{du_\infty}{dt}, (u_\ell - u_\infty)\varrho^2 \right\rangle dx_1 \right\} dt \\
= \int_0^t \left\langle \frac{d}{dt}(u_\ell - u_\infty)\varrho, (u_\ell - u_\infty)\varrho \right\rangle dt \qquad (9.48) \\
= \frac{1}{2} |(u_\ell - u_\infty)\varrho|_{2,\Omega_\ell}^2(t) \quad \text{a.e. } t \in (0,T).
\end{aligned}
$$

Thus going back to (9.40) we obtain by integration in t

$$
\begin{aligned}
\frac{1}{2}|(u_\ell - u_\infty)\varrho|_{2,\Omega_\ell}^2(t) + \int_0^t \int_{\Omega_\ell} A\nabla u_\ell \nabla\{(u_\ell - u_\infty)\varrho^2\} \, dx \, dt \\
= \int_0^t \int_{\Omega_\ell} A_{22} \nabla_{x_2} u_\infty \nabla_{x_2}\{(u_\ell - u_\infty)\varrho^2\} \, dx \, dt. \qquad (9.49)
\end{aligned}
$$

Recalling (9.24) and the fact that u_∞ is independent of X_1 we obtain

$$\frac{1}{2}|(u_\ell - u_\infty)\varrho|_{2,\Omega_\ell}^2(t) + \int_0^t \int_{\Omega_\ell} A_{11}\nabla_{X_1}(u_\ell - u_\infty)\nabla_{X_1}\{(u_\ell - u_\infty)\varrho^2\}\, dx\, dt$$

$$+ \int_0^t \int_{\Omega_\ell} A_{12}\nabla_{X_2}(u_\ell - u_\infty)\nabla_{X_1}\{(u_\ell - u_\infty)\varrho^2\}\, dx\, dt$$

$$+ \int_0^t \int_{\Omega_\ell} A_{21}\nabla_{X_1}(u_\ell - u_\infty)\nabla_{X_2}(u_\ell - u_\infty)\varrho^2\, dx\, dt \qquad (9.50)$$

$$+ \int_0^t \int_{\Omega_\ell} A_{22}\nabla_{X_2}(u_\ell - u_\infty)\nabla_{X_2}(u_\ell - u_\infty)\varrho^2\, dx\, dt$$

$$= -\int_0^t \int_{\Omega_\ell} A_{12}\nabla_{X_2}u_\infty \nabla_{X_1}\{(u_\ell - u_\infty)\varrho^2\}\, dx\, dt.$$

Since $A_{12}\nabla_{X_2}u_\infty$ is independent of X_1 we see easily that this last integral vanishes. Thus, we derive after having performed the derivatives in the directions X_1

$$\frac{1}{2}|(u_\ell - u_\infty)\varrho|_{2,\Omega_\ell}^2(t) + \int_0^t \int_{\Omega_\ell} A\nabla(u_\ell - u_\infty)\nabla(u_\ell - u_\infty)\varrho^2\, dx\, dt$$

$$= -\frac{2}{\ell_1}\int_0^t \int_{\Omega_\ell} A_{12}\nabla_{X_2}(u_\ell - u_\infty)\nabla_{X_1}\varrho\left(\frac{X_1}{\ell_1}\right)(u_\ell - u_\infty)\varrho\, dx\, dt \qquad (9.51)$$

$$-\frac{2}{\ell_1}\int_0^t \int_{\Omega_\ell} A_{11}\nabla_{X_1}(u_\ell - u_\infty)\nabla_{X_1}\varrho\left(\frac{X_1}{\ell_1}\right)(u_\ell - u_\infty)\varrho\, dx\, dt.$$

Using (9.22), (9.23), (9.38) we get for some constant C independent of ℓ_1, ℓ

$$\frac{1}{2}|(u_\ell - u_\infty)\varrho|_{2,\Omega_\ell}^2(t) + \lambda\int_0^t \int_{\Omega_\ell} |\nabla(u_\ell - u_\infty)|^2\varrho^2\, dx\, dt$$

$$\leq \frac{C}{\ell_1}\int_0^t \int_{\Omega_\ell} |\nabla(u_\ell - u_\infty)||(u_\ell - u_\infty)\varrho|\, dx\, dt.$$

Using now the Cauchy–Schwarz inequality and noticing that ϱ vanishes outside of Ω_{ℓ_1} we obtain

$$\frac{1}{2}|(u_\ell - u_\infty)\varrho|_{2,\Omega_{\ell_1}}^2(t) + \lambda\int_0^t \int_{\Omega_{\ell_1}} |\nabla(u_\ell - u_\infty)|^2\varrho^2\, dx\, dt$$

$$\leq \frac{C}{\ell_1}\left\{\int_0^t \int_{\Omega_{\ell_1}} |\nabla(u_\ell - u_\infty)|^2\, dx\, dt\right\}^{1/2} \qquad (9.52)$$

$$\left\{\int_0^t \int_{\Omega_{\ell_1}} (u_\ell - u_\infty)^2\varrho^2\, dx\, dt\right\}^{1/2}.$$

Applying Lemma 3.3 for $\ell = \ell_1$ we obtain

$$\frac{1}{2}|(u_\ell - u_\infty)\varrho|^2_{2,\Omega_{\ell_1}}(t) + \lambda \int_0^t \int_{\Omega_{\ell_1}} |\nabla(u_\ell - u_\infty)|^2 \varrho^2 \, dx \, dt$$

$$\leq \frac{C}{\ell_1}\left\{\int_0^t \int_{\Omega_{\ell_1}} |\nabla(u_\ell - u_\infty)|^2 \, dx \, dt\right\}^{1/2}\left\{\int_0^t \int_{\Omega_{\ell_1}} |\nabla(u_\ell - u_\infty)|^2 \varrho^2 \, dx \, dt\right\}^{1/2}$$

$$= \frac{\sqrt{\lambda} C}{\ell_1}\left\{\int_0^t \int_{\Omega_{\ell_1}} |\nabla(u_\ell - u_\infty)|^2 \, dx \, dt\right\}^{1/2}$$

$$\frac{1}{\sqrt{\lambda}}\left\{\int_0^t \int_{\Omega_{\ell_0}} |\nabla(u_\ell - u_\infty)|^2 \varrho^2 \, dx \, dt\right\}^{1/2}.$$

From this follows easily that

$$\frac{1}{2}|(u_\ell - u_\infty)\varrho|^2_{2,\Omega_{\ell_1}}(t) + \lambda \int_0^t \int_{\Omega_{\ell_1}} |\nabla(u_\ell - u_\infty)|^2 \varrho^2 \, dx \, dt$$

$$\leq \frac{C^2 \lambda}{\ell_1^2} \int_0^t \int_{\Omega_{\ell_1}} |\nabla(u_\ell - u_\infty)|^2 \, dx \, dt. \tag{9.53}$$

This clearly implies that

$$\lambda \int_0^t \int_{\Omega_{\ell_1}} |\nabla(u_\ell - u_\infty)|^2 \varrho^2 \, dx \, dt \leq \frac{C^2 \lambda}{\ell_1^2} \int_0^t \int_{\Omega_{\ell_1}} |\nabla(u_\ell - u_\infty)|^2 \, dx \, dt.$$

Since $\varrho = 1$ on $\Omega_{\ell_1/2}$ we derive for $t = T$

$$\int_0^T \int_{\Omega_{\ell_1/2}} |\nabla(u_\ell - u_\infty)|^2 \, dx \, dt \leq \frac{C^2}{\ell_1^2} \int_0^T \int_{\Omega_{\ell_1}} |\nabla(u_\ell - u_\infty)|^2 \, dx \, dt \tag{9.54}$$

for every $\ell_1 \leq \ell$. By k applications of this formula we obtain for every k and for some constant independent of ℓ

$$\int_0^T \int_{\Omega_{\ell/2^k}} |\nabla(u_\ell - u_\infty)|^2 \, dx \, dt \leq \frac{C}{\ell^{2k}} \int_0^T \int_{\Omega_\ell} |\nabla(u_\ell - u_\infty)|^2 \, dx \, dt. \tag{9.55}$$

Going back to (9.53) we obtain also

$$\frac{1}{2}|(u_\ell - u_\infty)|^2_{2,\Omega_{\ell/2^k}}(t) \leq \frac{C^2 \lambda}{\ell^2} 2^{k-1} \int_0^T \int_{\Omega_{\ell/2^{k-1}}} |\nabla(u_\ell - u_\infty)|^2 \, dx \, dt$$

$$\leq \frac{C}{\ell^{2k}} \int_0^T \int_{\Omega_\ell} |\nabla(u_\ell - u_\infty)|^2 \, dx \, dt \quad \text{a.e. } t \in (0, T). \tag{9.56}$$

Thus collecting (9.55), (9.56) we obtain for every k

$$|u_\ell - u_\infty|^2_{L^\infty(0,T;L^2(\Omega_{\ell/2^k}))} + \int_0^T \int_{\Omega_{\ell/2^k}} |\nabla(u_\ell - u_\infty)|^2 \, dx \, dt$$
$$\leq \frac{C}{\ell^{2k}} \int_0^T \int_{\Omega_\ell} |\nabla(u_\ell - u_\infty)|^2 \, dx \, dt. \tag{9.57}$$

By the estimate (9.30) we have for some constant independent of ℓ

$$\int_0^T \int_{\Omega_\ell} |\nabla u_\ell|^2 \, dx \, dt \leq C\ell^p. \tag{9.58}$$

Moreover it holds that

$$\int_0^T \int_{\Omega_\ell} |\nabla u_\infty|^2 \, dx \, dt = \int_0^T \int_{(-\ell,\ell)^p} \int_\omega |\nabla_{X_2} u_\infty|^2 \, dX_2 \, dX_1 \, dt \leq C\ell^p \tag{9.59}$$

where C is independent of ℓ.

Going back to (9.57) we have obtained

$$|u_\ell - u_\infty|_{L^\infty(0,T;L^2(\Omega_{\ell/2^k}))} + \int_0^T \int_{\Omega_{\ell/2^k}} |\nabla(u_\ell - u_\infty)|^2 \, dx \, dt \leq \frac{C}{\ell^{2k-p}}. \tag{9.60}$$

Thus, choosing $2k - p > 2r$ and ℓ so large that $\frac{\ell}{2^k} > \ell_0$ the result follows i.e., we have (9.28). This completes the proof of the theorem. $\qquad\square$

9.2. A nonlinear case

In this section we would like to address a nonlinear case. As we already noticed an important point here is to have uniqueness for the limit problem. So, we will consider the problem that we addressed in Section 8.3.

More precisely, Ω_ℓ being defined as in Section 9.1 let us denote by

$$a_{ij}(x,t,u), \quad i,j = 1,\ldots,n \tag{9.61}$$

Carathéodory functions defined on $\mathbb{R}^p \times \omega \times (0,T) \times \mathbb{R}$, i.e., functions such that

$$(x,t) \mapsto a_{ij}(x,t,u) \quad \text{is measurable } \forall u \in \mathbb{R}, \tag{9.62}$$
$$u \mapsto a_{ij}(x,t,u) \quad \text{is continuous} \quad \text{a.e. } (x,t) \in \mathbb{R}^p \times \omega \times (0,T). \tag{9.63}$$

Denoting by $A = (a_{ij})$ the matrix of the coefficients above, we will make an assumption similar to (9.24) and we will suppose that

$$A = A(x,t,u) = \begin{pmatrix} A_{11}(x,t,u) & A_{12}(X_2,t,u) \\ A_{21}(x,t,u) & A_{22}(X_2,t,u) \end{pmatrix} \tag{9.64}$$

i.e., if the above splitting is takes place at the level of the p-th line and p-th column we assume that the last $n - p$ columns of the matrix are independent of X_1. As usual we will further assume for some positive constant λ and Λ

$$A(x,t,u)\xi\xi \geq \lambda|\xi|^2 \quad \forall\,\xi \in \mathbb{R}^n,\ \forall\,u \in \mathbb{R},\ \text{a.e. } (x,t) \in \mathbb{R}^p \times \omega \times (0,T), \quad (9.65)$$

$$|A(x,t,u)| \leq \Lambda \quad \forall\,u \in \mathbb{R},\ \text{a.e. } (x,t) \in \mathbb{R}^p \times \omega \times (0,T),\ \forall\,i,j = 1,\ldots,n, \quad (9.66)$$

where $|A|$ denotes for instance the matrix norm of A subordinated to the euclidean norm in $\mathbb{R}^n$ – i.e., $|A| = \mathrm{Sup}_{|x|\leq 1}|Ax|$. In order to insure uniqueness of the limit problem we will further assume that

$$|A_{12}(\mathrm{X}_2,t,u) - A_{12}(\mathrm{X}_2,t,v)| \leq C\omega(|u-v|) \quad \forall\,u,v \in \mathbb{R},$$
$$\text{a.e. } (\mathrm{X}_2,t) \in \omega \times (0,T), \quad (9.67)$$

$$|A_{22}(\mathrm{X}_2,t,u) - A_{22}(\mathrm{X}_2,t,v)| \leq C\omega(|u-v|) \quad \forall\,u,v \in \mathbb{R},$$
$$\text{a.e. } (\mathrm{X}_2,t) \in \omega \times (0,T), \quad (9.68)$$

where ω is a nondecreasing positive function defined on $\mathbb{R}^+$ such that

$$\int_{0+} \frac{ds}{\omega^2(s)} = +\infty. \quad (9.69)$$

(ω denotes at the same time the section of our cylindrical domain and the modulus of continuity defined above. This should however not be source of confusion.)

Under the above assumptions it results from Theorem 8.10 that there exists $u_\ell,\ u_\infty$ solution to

$$\begin{cases} u_\ell \in H^1(0,T;H_0^1(\Omega_\ell),H^{-1}(\Omega_\ell)), \\ \dfrac{d}{dt}(u_\ell,v)_2 + \displaystyle\int_{\Omega_\ell} A(x,t,u_\ell)\nabla u_\ell \nabla v\,dx = \langle f,v\rangle \quad \text{in } \mathcal{D}'(0,T), \\ \qquad\qquad \forall\,v \in H_0^1(\Omega_\ell), \\ u_\ell(\cdot,0) = u_0, \end{cases} \quad (9.70)$$

and

$$\begin{cases} u_\infty \in H^1(0,T;H_0^1(\omega),H^{-1}(\omega)), \\ \dfrac{d}{dt}(u_\infty,v)_2 + \displaystyle\int_\omega A_{22}(x,t,u_\infty)\nabla_{\mathrm{X}_2}u_\infty \nabla_{\mathrm{X}_2}v\,dx = \langle f,v\rangle \quad \text{in } \mathcal{D}'(0,T), \\ \qquad\qquad \forall\,v \in H_0^1(\omega), \\ u_\infty(\cdot,0) = u_0 \end{cases} \quad (9.71)$$

for any $u_0 = u_0(\mathrm{X}_2) \in L^2(\omega)$ and any $f \in L^2(0,T;H^{-1}(\omega))$. Moreover, due to our assumptions (9.68)–(9.69) the solution u_∞ to (9.71) is unique (see Theorem 8.11). Thus we can now study the asymptotic behaviour of u_ℓ, transporting in the parabolic case the results that we already obtained in the elliptic case. For that, we will further assume that for some $\Lambda' > 0$ it holds that – for $i,j,k = 1,\ldots,n$

$$|\partial_{x_k}a_{ij}(x,t,u)| \leq \Lambda' \quad \forall\,u \in \mathbb{R},\ \text{a.e. } (x,t) \in \mathbb{R}^p \times \omega \times (0,T). \quad (9.72)$$

Then we can show:

THEOREM 9.4. *Under the above assumptions for*

$$u_0 \in L^2(\omega), \qquad f \in L^2(0,T; H^{-1}(\omega)) \tag{9.73}$$

consider u_ℓ, u_∞ *solution to* (9.70), (9.71). *Then for any* $\ell_0 > 0$, $\forall r > 0$ *there exists a constant* C *independent of* ℓ *such that*

$$|u_\ell - u_\infty|_{1,\Omega_{\ell_0} \times (0,T)} \le \frac{C}{\ell^r}. \tag{9.74}$$

$(|\cdot|_{1,K}$ *denotes the usual* L^1*-norm on* K.)

REMARK 9.5. Note that a priori we are not sure that u_ℓ is unique. However, the above estimate holds for any u_ℓ solution to (9.71).

Before giving the proof of Theorem 9.4 we would like to prove the following estimate for u_ℓ, namely

LEMMA 9.5. *Let* u_ℓ *be a solution to* (9.70). *There exists a constant* $C = C(\lambda, p)$ *independent of* ℓ *such that*

$$|u_\ell|^2_{L^\infty(0,T;L^2(\Omega_\ell))} + \int_0^T ||\nabla u_\ell||^2_{2,\Omega_\ell} \, dt \le C\ell^p \{|u_0|^2_{2,\omega} + |f|^2_{L^2(0,T;H^{-1}(\omega))}\}. \tag{9.75}$$

PROOF. We choose $v = u_\ell$ in (9.70). It follows that

$$\frac{1}{2}\frac{d}{dt}|u_\ell|^2_{2,\Omega_\ell} + \int_\Omega A(x,t,u_\ell)\nabla u_\ell \nabla u_\ell \, dx = \langle f, u_\ell \rangle. \tag{9.76}$$

Using the ellipticity condition (9.65) we derive

$$\frac{1}{2}\frac{d}{dt}|u_\ell|^2_{2,\Omega_\ell} + \lambda \int_{\Omega_\ell} |\nabla u_\ell|^2 \, dx \le |f|_{*,\Omega_\ell}||\nabla u_\ell||_{2,\Omega_\ell}$$

where in the above formula $|\cdot|_{*,\Omega_\ell}$ denotes the strong dual norm of $H^{-1}(\Omega_\ell)$ when $H_0^1(\Omega_\ell)$ is equipped with the L^2-norm of the gradient. The proof is then identical to the proof of Lemma 9.3 – see below (9.32). This completes the proof of the lemma. $\qquad\square$

We then turn to the proof of the theorem.

PROOF OF THEOREM 9.4. Let H_ε be the function defined by (8.73), (8.74). Let ϱ be a smooth function satisfying (9.37), (9.38). Clearly

$$H_\varepsilon(u_\ell - u_\infty)\varrho^2\left(\frac{x_1}{\ell_1}\right) \in H_0^1(\Omega_\ell) \tag{9.77}$$

for any $\ell_1 \le \ell$. We derive then from (9.70), (9.71) that

$$\left\langle \frac{du_\ell}{dt}, H_\varepsilon(u_\ell - u_\infty)\varrho^2 \right\rangle + \int_{\Omega_\ell} A(x,t,u_\ell)\nabla u_\ell \nabla(H_\varepsilon \varrho^2) \, dx = \langle f, H_\varepsilon \varrho^2 \rangle$$

$$= \int_{(-\ell,\ell)^p} \left\langle \frac{du_\infty}{dt}, H_\varepsilon(u_\ell - u_\infty)\varrho^2 \right\rangle \tag{9.78}$$

$$+ \int_{\Omega_\ell} A_{22}(x_2,t,u_\infty)\nabla_{x_2} u_\infty \nabla_{x_2}(H_\varepsilon \varrho^2) \, dx.$$

(When there is no confusion we denote $H_\varepsilon(u_\ell - u_\infty)\varrho^2$ simply by $H_\varepsilon\varrho^2$.) Let us set

$$K_\varepsilon(z) = \int_0^z H_\varepsilon(\xi)\, d\xi. \tag{9.79}$$

It is clear that for smooth functions it holds that

$$\frac{d}{dt}\int_{\Omega_\ell} K_\varepsilon(u_\ell - u_\infty)\varrho^2\, dx = \left\langle \frac{d}{dt}(u_\ell - u_\infty)\varrho, H_\varepsilon(u_\ell - u_\infty)\varrho \right\rangle. \tag{9.80}$$

By density this holds also for u_ℓ, u_∞ as above. Let us for simplicity denote $A_{ij}(x, t, u_\ell)$ and $A_{ij}(x, t, u_\infty)$ by $A_{ij}(u_\ell)$ and $A_{ij}(u_\infty)$. Then by (9.78) we obtain

$$\frac{d}{dt}\int_{\Omega_\ell} K_\varepsilon(u_\ell - u_\infty)\varrho^2\, dx$$

$$+ \int_{\Omega_\ell} A_{11}(u_\ell)\nabla_{X_1} u_\ell \nabla_{X_1}(H_\varepsilon\varrho^2)\, dx$$

$$+ \int_{\Omega_\ell} \{A_{12}(u_\ell)\nabla_{X_2} u_\ell - A_{12}(u_\infty)\nabla_{X_2} u_\infty\}\nabla_{X_1}(H_\varepsilon\varrho^2)\, dx$$

$$+ \int_{\Omega_\ell} A_{21}(u_\ell)\nabla_{X_1} u_\ell \nabla_{X_2}(H_\varepsilon\varrho^2)\, dx \tag{9.81}$$

$$+ \int_{\Omega_\ell} \{A_{22}(u_\ell)\nabla_{X_2} u_\ell - A_{22}(u_\infty)\nabla_{X_2} u_\infty\}\nabla_{X_2}(H_\varepsilon\varrho^2)\, dx$$

$$= -\int_{\Omega_\ell} A_{12}(u_\infty)\nabla_{X_2} u_\infty \nabla_{X_1}(H_\varepsilon\varrho^2)\, dx = 0.$$

(To see that this last integral vanishes it is enough to integrate in X_1 by noting that $A_{12}(u_\infty)\nabla_{X_2} u_\infty$ is independent of X_1.) Since ϱ^2 is independent of X_2 we obtain

$$E \overset{\text{Def}}{=} \frac{d}{dt}\int_{\Omega_\ell} K_\varepsilon(u_\ell - u_\infty)\varrho^2\, dx$$

$$+ \int_{\Omega_\ell} A_{11}(u_\ell)\nabla_{X_1} u_\ell \nabla_{X_1}\varrho^2 H_\varepsilon\, dx$$

$$+ \int_{\Omega_\ell} \{A_{12}(u_\ell)\nabla_{X_2} u_\ell - A_{12}(u_\infty)\nabla_{X_2} u_\infty\}\nabla_{X_1}\varrho^2 H_\varepsilon\, dx$$

$$= -\int_{\Omega_\ell} A_{11}(u_\ell)\nabla_{X_1} u_\ell \nabla_{X_1} H_\varepsilon\varrho^2\, dx \tag{9.82}$$

$$- \int_{\Omega_\ell} \{A_{12}(u_\ell)\nabla_{X_2} u_\ell - A_{12}(u_\infty)\nabla_{X_2}(u_\infty)\}\nabla_{X_1} H_\varepsilon\varrho^2\, dx$$

$$- \int_{\Omega_\ell} A_{21}(u_\ell)\nabla_{X_1} u_\ell \nabla_{X_2} H_\varepsilon\varrho^2\, dx$$

$$- \int_{\Omega_\ell} \{A_{22}(u_\ell)\nabla_{X_2} u_\ell - A_{22}(u_\infty)\nabla_{X_2} u_\infty\}\nabla_{X_2} H_\varepsilon\varrho^2\, dx.$$

Using the facts that

$$\nabla_{X_1} u_\ell = \nabla_{X_1}(u_\ell - u_\infty),$$
$$A_{i2}(u_\ell)\nabla_{X_2} u_\ell - A_{i2}(u_\infty)\nabla_{X_2} u_\infty$$
$$= A_{i2}(u_\ell)\nabla_{X_2}(u_\ell - u_\infty) + \{A_{i2}(u_\ell) - A_{i2}(u_\infty)\}\nabla_{X_2} u_\infty$$

we obtain

$$E = -\int_{\Omega_\ell} A(u_\ell)\nabla(u_\ell - u_\infty)\nabla H_\varepsilon \varrho^2$$

$$-\sum_{i=1}^{2}\int_{\Omega_\ell} \{A_{i2}(u_\ell) - A_{i2}(u_\infty)\}\nabla_{X_2} u_\infty \nabla_{X_i} H_\varepsilon \varrho^2 \, dx. \tag{9.83}$$

By the ellipticity condition and (9.67), (9.68) we obtain, setting $S_\varepsilon = \{x \in \Omega_\ell \mid (u_\ell - u_\infty)(x) > \varepsilon\}$,

$$E \le -\int_{S_\varepsilon} A(u_\ell)\nabla(u_\ell - u_\infty)\nabla(u_\ell - u_\infty)H'_\varepsilon \varrho^2$$

$$+ C\int_{S_\varepsilon} \omega(u_\ell - u_\infty)|\nabla_{X_2} u_\infty||\nabla(u_\ell - u_\infty)|H'_\varepsilon \varrho^2 \, dx \tag{9.84}$$

$$\le \frac{1}{I_\varepsilon}\left\{-\lambda\int_{S_\varepsilon} \frac{|\nabla(u_\ell - u_\infty)|^2}{\omega^2(u_\ell - u_\infty)}\varrho^2 \, dx + C\int_{S_\varepsilon} \frac{|\nabla(u_\ell - u_\infty)|}{\omega(u_\ell - u_\infty)}|\nabla_{X_2} u_\infty|\varrho^2 \, dx\right\}$$

for some constant C independent of ε. (Recall the Definition (8.73) of H_ε in order to differentiate it).

Applying now the Young inequality

$$ab \le \frac{\lambda}{2C}a^2 + \frac{C}{2\lambda}b^2$$

we derive that

$$C\int_{S_\varepsilon} \frac{|\nabla(u_\ell - u_\infty)|}{\omega(u_\ell - u_\infty)}|\nabla_{X_2} u_\infty|\varrho^2 \, dx$$

$$\le \frac{\lambda}{2}\int_{S_\varepsilon} \frac{|\nabla(u_\ell - u_\infty)|^2}{\omega^2(u_\ell - u_\infty)}\varrho^2 \, dx + \frac{C^2}{2\lambda}\int_{S_\varepsilon} |\nabla_{X_2} u_\infty|^2\varrho^2 \, dx$$

and thus by (9.84)

$$E \le \frac{1}{I_\varepsilon}\frac{C^2}{2\lambda}\int_{S_\varepsilon} |\nabla_{X_2} u_\infty|^2\varrho^2 \, dx. \tag{9.85}$$

Recalling the definition of E (see (9.82)) and integrating in t we obtain

$$\int_{\Omega_\ell} K_\varepsilon(u_\ell - u_\infty)\varrho^2 \, dx$$

$$+ \int_0^t \int_{\Omega_\ell} A_{11}(u_\ell)\nabla_{X_1} u_\ell \nabla_{X_1} \varrho^2 H_\varepsilon \, dx \, dt$$

$$+ \int_0^t \int_{\Omega_\ell} \{A_{12}(u_\ell)\nabla_{X_2} u_\ell - A_{12}(u_\infty)\nabla_{X_2} u_\infty\}\nabla_{X_1} \varrho^2 H_\varepsilon \, dx \, dt$$

$$\leq \frac{C^2}{2\lambda I_\varepsilon} \int_0^t \int_{\Omega_\ell} |\nabla_{X_2} u_\infty|^2 \varrho^2 \, dx \, dt.$$

(9.86)

Noting that when $\varepsilon \to 0$

$$I_\varepsilon \to +\infty, \qquad H_\varepsilon \to \chi_{\mathbb{R}^+}, \qquad K_\varepsilon(z) \to z^+$$

we derive by taking the limit in ε in (9.86) that

$$\int_{[u_\ell - u_\infty > 0]} (u_\ell - u_\infty)^+ \varrho^2 \, dx$$

$$+ \int_0^t \int_{[u_\ell - u_\infty > 0]} A_{11}(u_\ell)\nabla_{X_1} u_\ell \nabla_{X_1} \varrho^2 \, dx \, dt$$

$$+ \int_0^t \int_{[u_\ell - u_\infty > 0]} \{A_{12}(u_\ell)\nabla_{X_2} u_\ell - A_{12}(u_\infty)\nabla_{X_2} u_\infty\}\nabla_{X_1} \varrho^2 \, dx \, dt$$

$$\leq 0.$$

(9.87)

$([u_\ell - u_\infty > 0] = \{x \in \Omega_\ell \mid (u_\ell - u_\infty)(x) > 0\}.)$

To exploit this inequality we introduce the functions

$$\tilde{a}_{ij}(x, t, z) = \int_0^z a_{ij}(x, t, s) \, ds.$$

(9.88)

For $u \in H^1(\Omega_\ell)$ it holds that

$$\tilde{a}_{ij}(x, t, u) \in H^1(\Omega_\ell) \quad \text{a.e. } t \in (0, T)$$

and for every $k = 1, \ldots, n$ it holds that

$$\partial_{x_k} \tilde{a}_{ij}(x, t, u) = a_{ij}(x, t, u)\partial_{x_k} u + \int_0^u \partial_{x_k} a_{ij}(x, t, s) \, ds.$$

(9.89)

This remark being made, if we denote by E' the sum of the last two integrals in (9.87) we obtain, dropping for simplicity the dependence in t, x of the coefficients

$$E' = \int_0^t \sum_{i,j=1}^p \int_{[u_\ell - u_\infty > 0]} a_{ij}(u_\ell)\partial_{x_j} u_\ell \partial_{x_i} \varrho^2 \, dx \, dt$$

$$+ \int_0^t \sum_{i=1}^p \sum_{j=p+1}^n \int_{[u_\ell - u_\infty > 0]} \{a_{ij}(u_\ell)\partial_{x_j} u_\ell - a_{ij}(u_\infty)\partial_{x_j} u_\infty\}\partial_{x_i} \varrho^2 \, dx \, dt.$$

Thus by (9.89) we obtain

$$
E' = \int_0^t \sum_{i,j=1}^p \int_{[u_\ell - u_\infty > 0]} \left\{ \partial_{x_j} \tilde{a}_{ij}(u_\ell) - \int_0^{u_\ell} \partial_{x_j} a_{ij}(s) \, ds \right\} \partial_{x_i} \varrho^2 \, dx \, dt
$$

$$
+ \int_0^t \sum_{i=1}^p \sum_{j=p+1}^n \int_{[u_\ell - u_\infty > 0]} \left\{ \partial_{x_j} \tilde{a}_{ij}(u_\ell) - \partial_{x_j} \tilde{a}_{ij}(u_\infty) \right.
$$

$$
\left. - \int_{u_\infty}^{u_\ell} \partial_{x_j} a_{ij}(s) \, ds \right\} \partial_{x_i} \varrho^2 \, dx \, dt.
$$

From (9.89) we derive

$$
\partial_{x_k} \tilde{a}_{ij}(u_\infty) = \int_0^{u_\infty} \partial_{x_k} a_{ij}(x,t,s) \, ds \quad \forall k = 1, \dots, p
$$

so that the above expression of E' can be written

$$
E' = \int_0^t \sum_{i=1}^p \sum_{j=1}^n \int_{[u_\ell - u_\infty > 0]} \left[\partial_{x_j} (\tilde{a}_{ij}(u_\ell) - \tilde{a}_{ij}(u_\infty)) \partial_{x_i} \varrho^2 \right.
$$

$$
\left. - \int_{u_\infty}^{u_\ell} \partial_{x_j} a_{ij}(s) \, ds \partial_{x_i} \varrho^2 \right] dx \, dt. \tag{9.90}
$$

Introducing $w = (u_\ell - u_\infty)^+$ this leads to

$$
E' = \int_0^t \sum_{i=1}^p \sum_{j=1}^n \int_{\Omega_\ell} \left[\partial_{x_j} (\tilde{a}_{ij}(u_\infty + w) - \tilde{a}_{ij}(u_\infty)) \partial_{x_i} \varrho^2 \right.
$$

$$
\left. - \int_{u_\infty}^{u_\infty + w} \partial_{x_j} a_{ij}(s) \, ds \partial_{x_i} \varrho^2 \right] dx \, dt.
$$

Integrating by parts we obtain

$$
E' = \int_0^t \sum_{i=1}^p \sum_{j=1}^n - \int_{\Omega_\ell} \left[\tilde{a}_{ij}(u_\infty + w) - \tilde{a}_{ij}(u_\infty) \partial_{x_i x_j}^2 \varrho^2 \right.
$$

$$
\left. + \int_{u_\infty}^{u_\infty + w} \partial_{x_j} a_{ij}(s) \, ds \partial_{x_i} \varrho^2 \right] dx \, dt
$$

$$
= \int_0^t \sum_{i=1}^p \sum_{j=1}^n - \int_{\Omega_\ell} \left[\int_{u_\infty}^{u_\infty + w} a_{ij}(s) \, ds \partial_{x_i x_j}^2 \varrho^2 \right. \tag{9.91}
$$

$$
\left. + \int_{u_\infty}^{u_\infty + w} \partial_{x_j} a_{ij}(s) \, ds \partial_{x_i} \varrho^2 \right] dx \, dt.
$$

Since the functions a_{ij} and $\partial_{x_k} a_{ij}$ are bounded and since it holds that

$$
\left| \partial_{x_i x_j}^2 \varrho^2 \left(\frac{x_1}{\ell_1} \right) \right|, \left| \partial_{x_i} \varrho^2 \left(\frac{x_1}{\ell_1} \right) \right| \le \frac{C}{\ell_1}
$$

where C is a constant independent of ℓ_1 and ℓ we get – note also that it is enough to integrate on Ω_{ℓ_1} –

$$|E'| \leq \frac{C}{\ell_1} \int_0^t \int_{\Omega_{\ell_1}} w \, dx \, dt = \frac{C}{\ell_1} \int_0^t \int_{\Omega_{\ell_1}} (u_\ell - u_\infty)^+ \, dx \, dt. \qquad (9.92)$$

Recalling the definition of E' and going back to (9.87) we get

$$\int_{\Omega_{\ell_1}} (u_\ell - u_\infty)^+ \varrho^2 \, dx \leq \frac{C}{\ell_1} \int_0^t \int_{\Omega_{\ell_1}} (u_\ell - u_\infty)^+ \, dx \, dt$$

$$\leq \frac{C}{\ell_1} \int_0^T \int_{\Omega_{\ell_1}} (u_\ell - u_\infty)^+ \, dx \, dt. \qquad (9.93)$$

Since $\varrho = 1$ on $\Omega_{\ell_1/2}$ this gives after integration in T

$$\int_0^T \int_{\Omega_{\ell_1/2}} (u_\ell - u_\infty)^+ \, dx \, dt \leq \frac{CT}{\ell_1} \int_0^T \int_{\Omega_{\ell_1}} (u_\ell - u_\infty)^+ \, dx \, dt \quad \forall \ell_1 \leq \ell. \qquad (9.94)$$

Since $-u_\ell$, $-u_\infty$ satisfy similar equations we obtain with C independent of $\ell_1 \leq \ell$

$$\int_0^T \int_{\Omega_{\ell_1/2}} |u_\ell - u_\infty| \, dx \, dt \leq \frac{C}{\ell_1} \int_0^T \int_{\Omega_{\ell_1}} |u_\ell - u_\infty| \, dx \, dt \quad \forall \ell_1 \leq \ell.$$

Iterating this inequality we get

$$\int_0^T \int_{\Omega_{\ell/2^k}} |u_\ell - u_\infty| \, dx \, dt \leq \frac{C}{\ell^k} \int_0^T \int_{\Omega_\ell} |u_\ell - u_\infty| \, dx \, dt$$

$$\leq \frac{C}{\ell^k} \int_0^T |u_\ell - u_\infty|_{2,\Omega_\ell} |\Omega_\ell|^{1/2} \, dt$$

$$= \frac{C}{\ell^{k-p/2}} \int_0^T |u_\ell - u_\infty|_{2,\Omega_\ell} \, dt$$

(by the Cauchy–Schwarz inequality). Using in this last integral the Poincaré inequality we derive

$$\int_0^T \int_{\Omega_{\ell/2^k}} |u_\ell - u_\infty| \, dx \, dt \leq \frac{C}{\ell^{k-p/2}} \int_0^T ||\nabla(u_\ell - u_\infty)||_{2,\Omega_\ell} \, dt.$$

By Lemma 9.5 this gives

$$\int_0^T \int_{\Omega_{\ell/2^k}} |u_\ell - u_\infty| \, dx \, dt \leq \frac{C}{\ell^{k-p}}.$$

Choosing $k - p > r$ and then ℓ such that $\frac{\ell}{2^k} > \ell_0$ the result follows. This completes the proof of the theorem. □

As we did in the elliptic case, it is also possible to prove that u_ℓ converges to u_∞ in spaces more adapted to the problem. We will suppose here that

$$f \in L^2((0, T) \times \omega). \qquad (9.95)$$

Then we have

THEOREM 9.6. *Let us assume the conditions of Theorem 9.4. Moreover let us suppose that there exists ω a positive, nondecreasing function satisfying (9.69) and a constant C such that*

$$|a_{ij}(x,t,u) - a_{ij}(x,t,v)| \leq C\omega(|u-v|) \quad \text{a.e. } (x,t) \in \mathbb{R}^P \times \omega \times (0,T),$$
$$\forall u,v \in \mathbb{R}, \quad \forall i,j = 1,\ldots,n. \tag{9.96}$$

Then, for f satisfying (9.95), for every $\ell_0 > 0$, it holds that

$$u_\ell(t) \to u_\infty(t) \quad \text{in } L^2(\Omega_{\ell_0}) \text{ a.e. } t \in (0,T), \tag{9.97}$$

$$u_\ell \to u_\infty \quad \text{in } L^2(0,T;H^1(\Omega_{\ell_0})) \tag{9.98}$$

when $\ell \to +\infty$.

PROOF. Note that under the assumptions (9.96) the solution u_ℓ to (9.70) is unique. We start by a pointwise estimate of u_ℓ. For that we introduce u_∞^+ and u_∞^- the solution to

$$\begin{cases} u_\infty^+ \in H^1(0,T;H_0^1(\omega),H^{-1}(\omega)), \\ \dfrac{d}{dt}(u_\infty^+,v)_2 + \displaystyle\int_\omega A_{22}(\mathrm{x}_2,t,u_\infty^+)\nabla_{\mathrm{X}_2}u_\infty^+\nabla_{\mathrm{X}_2}v\,d\mathrm{x}_2 \\ \qquad = \displaystyle\int_\omega f^+ v\,d\mathrm{x}_2 \quad \text{in } \mathcal{D}'(0,T), \; \forall v \in H_0^1(\omega), \\ u_\infty^+(\cdot,0) = u_0^+, \end{cases} \tag{9.99}$$

$$\begin{cases} u_\infty^- \in H^1(0,T;H_0^1(\omega),H^{-1}(\omega)), \\ \dfrac{d}{dt}(u_\infty^-,v)_2 + \displaystyle\int_\omega A_{22}(\mathrm{x}_2,t,u_\infty^-)\nabla_{\mathrm{X}_2}u_\infty^-\nabla_{\mathrm{X}_2}v\,d\mathrm{x}_2 \\ \qquad = -\displaystyle\int_\omega f^- v\,d\mathrm{x}_2 \quad \text{in } \mathcal{D}'(0,T), \; \forall v \in H_0^1(\omega), \\ u_\infty^-(\cdot,0) = -u_0^-. \end{cases} \tag{9.100}$$

$((\cdot)^+$ and $(\cdot)^-$ denote respectively the positive and negative parts of functions.) We claim that u_∞^+ and u_∞^- are upper and subsolution to problem (9.70). Let us show for instance that u_∞^+ is an uppersolution. For this purpose set

$$\Gamma_0 = (-\ell,\ell)^P \times \partial\omega, \qquad V = H_0^1(\Omega_\ell;\Gamma_0). \tag{9.101}$$

We claim first that for $v \in H^1(0,T;H_0^1(\omega),H^{-1}(\omega))$ its canonical extension belongs to $H^1(0,T;V,V')$. Indeed this extension $\tilde{v}$ belongs clearly to $L^2(0,T;V)$. Moreover, for any $w \in V$ it holds that $w(\mathrm{x}_1,\cdot) \in H_0^1(\omega)$ a.e. $\mathrm{x}_1 \in (-\ell,\ell)^P$ and thus

$$\int_0^T \left\langle \frac{dv}{dt}, w(\mathrm{x}_1,\cdot) \right\rangle \varphi\,dt = -\int_0^T \int_\omega v(\mathrm{x}_2)w(\mathrm{x}_1,\mathrm{x}_2)\varphi'(t)\,d\mathrm{x}_2\,dt.$$

Integrating this equality on $(-\ell, \ell)^p$ we get – see (9.5):

$$\int_0^T \left\langle \frac{\widetilde{dv}}{dt}, w \right\rangle \varphi \, dt = - \int_0^T \int_{\Omega_\ell} v w \varphi' \, dx \, dt \quad \forall \varphi \in \mathcal{D}(0, T).$$

Thus

$$\frac{dv}{dt} = \frac{\widetilde{dv}}{dt} \in L^2(0, T, V').$$

This shows that

$$u_\infty^+ \in H^1(0, T; V, V').$$

Moreover, for $v \in H_0^1(\Omega_\ell), v(\mathbf{x}_1, \cdot) \in H_0^1(\omega)$ for a.e. $\mathbf{x}_1 \in (-\ell, \ell)^p$ and from (9.99) we derive

$$\frac{d}{dt}(u_\infty^+, v(\mathbf{x}_1, \cdot))_2 + \int_\omega A_{22}(\mathbf{x}_2, t, u_\infty^+) \nabla_{\mathbf{x}_2} u_\infty^+ \nabla_{\mathbf{x}_2} v(\mathbf{x}_1, \cdot) \, d\mathbf{x}_2$$

$$= \int_\omega f^+ v(\mathbf{x}_1, \cdot) \, d\mathbf{x}_2 \quad \text{in } \mathcal{D}'(0, T).$$

Integrating in $\mathbf{x}_1$ it follows that

$$\frac{d}{dt}(u_\infty^+, v)_2 + \int_{\Omega_\ell} A(x, t, u_\infty^+) \nabla u_\infty^+ \nabla v \, dx = \int_{\Omega_\ell} f^+ v \, dx \quad \text{in } \mathcal{D}'(0, T). \quad (9.102)$$

(Note that $\nabla_{\mathbf{x}_1} u_\infty^+ = 0$ and

$$\int_{\Omega_\ell} A_{12}(\mathbf{x}_2, t, u_\infty^+) \nabla_{\mathbf{x}_2} u_\infty^+ \nabla_{\mathbf{x}_1} v \, dx = 0.)$$

Since $f^+ \geq f$, for $v \in H_0^1(\Omega_\ell)$, $v \geq 0$ we derive from (9.102) that

$$\frac{d}{dt}(u_\infty^+, v)_2 + \int_{\Omega_\ell} A(x, t, u_\infty^+) \nabla u_\infty^+ \nabla v \, dx \geq \frac{d}{dt}(u_\ell, v)_2 + \int_{\Omega_\ell} A(x, t, u_\ell) \nabla u_\ell \nabla v \, dx.$$

Using the comparison principle of [10] – see also Remark 8.4 – we can show that

$$u_\ell \leq u_\infty^+ \quad \text{in } \Omega_\ell \times (0, T).$$

Arguing the same way for u_∞^- we arrive at

$$u_\infty^- \leq u_\ell \leq u_\infty^+ \quad \text{a.e. in } \Omega_\ell \times (0, T). \quad (9.103)$$

We can now derive an a priori estimate for u_ℓ. Indeed, as usual, one can estimate u_ℓ in function of its L^2-norm, then (9.103) will provide us with a bound for u_ℓ. Let us develop that more precisely. For this purpose consider ϱ a smooth function of $\mathbf{x}_1$ only such that

$$0 \leq \varrho \leq 1, \qquad \varrho = 1 \quad \text{on } (-\ell_0, \ell_0)^p, \qquad \varrho = 0 \quad \text{outside } (-\ell_0 - 1, \ell_0 + 1)^p. \quad (9.104)$$

Taking $v = u_\ell \varrho^2$ in (9.70) we obtain

$$\left\langle \frac{du_\ell}{dt}, u_\ell \varrho^2 \right\rangle + \int_{\Omega_\ell} A \nabla u_\ell \nabla (u_\ell \varrho^2) \, dx = \int_{\Omega_\ell} f u_\ell \varrho^2 \, dx$$

$$\Leftrightarrow \quad \left\langle \frac{du_\ell}{dt}, u_\ell \varrho^2 \right\rangle + \int_{\Omega_\ell} A \nabla u_\ell \nabla u_\ell \varrho^2 \, dx = \int_{\Omega_\ell} f u_\ell \varrho^2 \, dx - \int_{\Omega_\ell} A \nabla u_\ell \nabla \varrho^2 u_\ell \, dx.$$

By (9.65), (9.66) we obtain, since ϱ is smooth,

$$\frac{1}{2} \frac{d}{dt} |u_\ell \varrho|^2_{2,\Omega_\ell} + \lambda \int_{\Omega_\ell} |\nabla u_\ell|^2 \varrho^2 \, dx \le |f|_{2,\Omega_{\ell_0+1}} |u_\ell|_{2,\Omega_{\ell_0+1}} + C \int_{\Omega_\ell} |\nabla u_\ell| |u_\ell| \varrho \, dx$$

$$\le |f|_{2,\Omega_{\ell_0+1}} |u_\ell|_{2,\Omega_{\ell_0+1}} + C ||\nabla u_\ell| \varrho|_{2,\Omega_\ell} |u_\ell|_{2,\Omega_{\ell_0+1}}$$

for some constant C. Using Young's inequality in the last term we get

$$C ||\nabla u_\ell| \varrho|_{2,\Omega_\ell} |u_\ell|_{2,\Omega_{\ell_0+1}} \le \frac{\lambda}{2} ||\nabla u_\ell| \varrho|^2_{2,\Omega_\ell} + \frac{C^2}{2\lambda} |u_\ell|^2_{2,\Omega_{\ell_0+1}}$$

and thus, for some constant C,

$$\frac{d}{dt} |u_\ell \varrho|^2_{2,\Omega_\ell} + \lambda ||\nabla u_\ell| \varrho|^2_{2,\Omega_\ell} \le C (|f|_{2,\Omega_{\ell_0+1}} + |u_\ell|_{2,\Omega_{\ell_0+1}}) |u_\ell|_{2,\Omega_{\ell_0+1}}$$

$$\le C(\ell_0)$$

thanks to (9.103). Integrating in t we obtain

$$|\varrho u_\ell|_{L^2(0,T;H^1(\Omega_\ell))} \le C \tag{9.105}$$

where C is a constant independent of ℓ. Using now (9.70) we have for any $v \in H_0^1(\Omega_\ell)$

$$\left| \left\langle \frac{du_\ell}{dt}, \varrho v \right\rangle \right| = \left| - \int_{\Omega_\ell} A \nabla u_\ell \nabla (\varrho v) \, dx + \int_{\Omega_\ell} f \varrho v \, dx \right|$$

$$\le C \left\{ ||\nabla u_\ell||_{2,\Omega_{\ell_0+1}} + |f|_{2,\Omega_{\ell_0+1}} \right\} ||\nabla v||_{2,\Omega_\ell}.$$

Thus if we denote by $|\cdot|_{*,\Omega_\ell}$ the strong dual norm in $H^{-1}(\Omega_\ell)$ we get

$$\left| \frac{d}{dt} (\varrho u_\ell) \right|_{*,\Omega_\ell} \le C \left\{ ||\nabla u_\ell||_{2,\Omega_{\ell_0+1}} + |f|_{2,\Omega_{\ell_0+1}} \right\}.$$

Taking into account (9.105), with perhaps $\varrho = 1$ on $(-\ell_0 - 1, \ell_0 + 1)^p$ instead of $(-\ell_0, \ell_0)^p$, we derive

$$\left| \frac{d}{dt} (\varrho u_\ell) \right|_{L^2(0,T;H^{-1}(\Omega_\ell))} \le C. \tag{9.106}$$

Thus we know now that ϱu_ℓ is bounded in $H^1(0,T; H_0^1(\Omega_{\ell_0+1}), H^{-1}(\Omega_{\ell_0+1}))$ and thus – up to a subsequence – we can assume that for some u it holds that

$$\varrho u_\ell \rightharpoonup u \quad \text{in } L^2(0,T; H_0^1(\Omega_{\ell_0+1})), \qquad \varrho u_\ell \to u \quad \text{in } L^2(0,T; L^2(\Omega_{\ell_0+1})). \tag{9.107}$$

Since by (9.74) we have also

$$u_\ell \to u_\infty \quad \text{in } L^1(0,T; L^1(\Omega_{\ell_0+1}))$$

we have clearly

$$\varrho u_\ell \rightharpoonup \varrho u_\infty \quad \text{in } L^2(0,T; H_0^1(\Omega_{\ell_0+1})),$$

$$\varrho u_\ell \to \varrho u_\infty \quad \text{in } L^2(0,T; L^2(\Omega_{\ell_0+1})) \tag{9.108}$$

and since the limit value is unique, this is the whole sequence ϱu_ℓ that satisfies (9.108). Taking $v = u_\ell \varrho^2$ in (9.70), (9.71) and integrating in t we obtain with obvious notation

$$\frac{1}{2}|u_\ell(t)\varrho|_{2,\Omega_\ell}^2 + \int_0^t \int_{\Omega_\ell} A(x,t,u_\ell)\nabla u_\ell \nabla u_\ell \varrho^2 \, dx \, dt$$

$$= \frac{1}{2}|u_0\varrho|_{2,\Omega_\ell}^2 + \int_0^t \left\langle \frac{\widetilde{du_\infty}}{dt}, u_\ell \varrho^2 \right\rangle dt$$

$$+ \int_0^t \int_{\Omega_\ell} A_{22}(x,t,u_\infty)\nabla_{X_2} u_\infty \nabla_{X_2} u_\ell \varrho^2 \, dx \, dt \tag{9.109}$$

$$- \int_0^t \int_{\Omega_\ell} A(x,t,u_\ell)\nabla u_\ell \nabla \varrho^2 u_\ell \, dx \, dt.$$

From (9.108) we deduce easily that

$$\lim_{\ell \to +\infty} \int_0^t \left\langle \frac{\widetilde{du_\infty}}{dt}, u_\ell \varrho^2 \right\rangle dt$$

$$= \int_0^t \left\langle \frac{\widetilde{du_\infty}}{dt}, u_\infty \varrho^2 \right\rangle \tag{9.110}$$

$$= \frac{1}{2}|u_\infty(t)\varrho|_{2,\Omega_\ell}^2 - \frac{1}{2}|u_0\varrho|_{2,\Omega_\ell}^2,$$

$$\lim_{\ell \to +\infty} \int_0^t \int_{\Omega_\ell} A_{22}(x,t,u_\infty)\nabla_{X_2} u_\infty \nabla_{X_2} u_\ell \varrho^2 \, dx \, dt$$

$$= \int_0^t \int_{\Omega_\ell} A_{22}(x,t,u_\infty)\nabla_{X_2} u_\infty \nabla_{X_2} u_\infty \varrho^2 \, dx \, dt. \tag{9.111}$$

Moreover, we claim that

$$A(x,t,u_\ell)\nabla \varrho^2 u_\ell \to A(x,t,u_\infty)\nabla \varrho^2 u_\infty \quad \text{in } \mathbb{L}^2(0,T; L^2(\Omega_{\ell_0+1})). \tag{9.112}$$

Recall that $\mathbb{L}^2(0,T; L^2(\Omega_{\ell_0+1})) = L^2(0,T; L^2(\Omega_{\ell_0+1}))^n$, and that these functions are vanishing out of Ω_{ℓ_0+1}. Indeed, if (9.112) fails, then for a subsequence still labeled by ℓ it holds that

$$|A(x,t,u_\ell)\nabla \varrho^2 u_\ell - A(x,t,u_\infty)\nabla \varrho^2 u_\infty|_{L^2(0,T,L^2(\Omega_{\ell_0+1}))} \geq \varepsilon. \tag{9.113}$$

But since $u_\ell \to u_\infty$ in $L^1(0,T; L^1(\Omega_{\ell_0+1}))$ for a subsequence it holds that

$$A(x,t,u_\ell)\nabla \varrho^2 u_\ell \to A(x,t,u_\infty)\nabla \varrho^2 u_\infty \quad \text{a.e. on } \Omega_{\ell_0+1} \times (0,T).$$

But then by the Lebesgue theorem and (9.103) it follows that

$$A(x, t, u_\ell)\nabla \varrho^2 u_\ell \to A(x, t, u_\infty)\nabla \varrho^2 u_\infty \quad \text{in } \mathbb{L}^2(0, T; L^2(\Omega_{\ell_0+1}))$$

which contradicts (9.113). Combining then (9.109)–(9.112) we obtain

$$
\lim_{\ell \to +\infty} \frac{1}{2}|u_\ell(t)\varrho|^2_{2,\Omega_\ell} + \int_0^t \int_{\Omega_\ell} A(x, t, u_\ell)\nabla u_\ell \nabla u_\ell \varrho^2 \, dx \, dt
$$
$$
= \frac{1}{2}|u_\infty(t)\varrho|^2_{2,\Omega_\ell} + \int_0^t \int_{\Omega_\ell} A_{22}(x, t, u_\infty)\nabla_{X_2} u_\infty \nabla_{X_2} u_\infty \varrho^2 \, dx \, dt \quad (9.114)
$$
$$
- \int_0^t \int_{\Omega_\ell} A(x, t, u_\infty)\nabla u_\infty \nabla \varrho^2 u_\infty \, dx \, dt.
$$

Using the ellipticity condition and the fact that

$$u_\ell(0) = u_\infty(0)$$

we obtain for $t \in (0, T)$

$$
\frac{1}{2}|(u_\ell - u_\infty)(t)\varrho|^2_{2,\Omega_\ell} + \lambda \int_0^t \||\nabla(u_\ell - u_\infty)|\varrho|^2_{2,\Omega_\ell} \, dt
$$
$$
\leq \int_0^t \left\langle \frac{d}{dt}(u_\ell - u_\infty)\varrho, (u_\ell - u_\infty)\varrho \right\rangle dt
$$
$$
+ \int_0^t \int_{\Omega_\ell} A(x, t, u_\ell)\nabla(u_\ell - u_\infty)\nabla(u_\ell - u_\infty)\varrho^2 \, dx \, dt
$$
$$
= -|u_0\varrho|^2_{2,\Omega_\ell} + \frac{1}{2}|u_\ell(t)\varrho|^2_{2,\Omega_\ell} + \frac{1}{2}|u_\infty(t)\varrho|^2_{2,\Omega_\ell} - \int_0^t \left\langle \frac{du_\ell}{dt}, u_\infty \varrho^2 \right\rangle dt
$$
$$
- \int_0^t \left\langle \frac{\widetilde{du_\infty}}{dt}, u_\ell \varrho^2 \right\rangle dt + \int_0^t \int_{\Omega_\ell} A(x, t, u_\ell)\nabla u_\ell \nabla u_\ell \varrho^2 \, dx \, dt \quad (9.115)
$$
$$
+ \int_0^t \int_{\Omega_\ell} A(x, t, u_\ell)\nabla u_\infty \nabla u_\infty \varrho^2 \, dx \, dt
$$
$$
- \int_0^t \int_{\Omega_\ell} A(x, t, u_\ell)\nabla u_\infty \nabla u_\ell \varrho^2 \, dx \, dt
$$
$$
- \int_0^t \int_{\Omega_\ell} A(x, t, u_\ell)\nabla u_\ell \nabla u_\infty \varrho^2 \, dx \, dt.
$$

Using the same arguments as above and (9.114) we obtain

$$
\lim_{\ell \to \infty} |(u_\ell - u_\infty)(t)|^2_{2,\Omega_\ell} + \lambda \int_0^t \||\nabla u_\ell - u_\infty)|\varrho|^2_{2,\Omega_\ell} \, dt = 0 \quad t \in (0, T).
$$

This completes the proof of the theorem. $\qquad\qquad\qquad\square$

Open problems

1. It would be interesting to see in Theorem 9.2 when it is possible to obtain an exponential rate of convergence.

2. A theory of convergence in higher Sobolev spaces in the spirit of Section 3.3 has yet to be developed.

3. We do not know if the results of Section 9.2 extend for more general boundary conditions.

4. Could it be possible to replace the assumption (9.95) in Theorem 9.6 by $f \in L^2(0, T; H^{-1}(\omega))$?

5. We do not know the speed of convergence in Theorem 9.6.

Concluding Remark

We have shown how relatively simple techniques allow us to predict the asymptotic behaviour of problems set in cylindrical domains which become unbounded. Of course, we have restricted ourselves to a small number of such problems. Almost any problem of mathematical physics can be addressed in these terms.

Among the most important topics that we left out are the calculus of variations (see for instance [32] for an example), free-boundary problems, higher order problems and problems with periodic or other boundary conditions to quote only a few. We hope that these gaps will be filled in the near future, and that this will lead to further new techniques.

Bibliography

1. R.A. Adams, *Sobolev spaces*, Acad. Press, New York – San Francisco – London, 1975.
2. H. Amann, *Linear and quasilinear parabolic problems*, Abstract Linear Theory, vol. 1, Birkhäuser, 1995.
3. N. André and M. Chipot, *A remark on uniqueness for quasilinear elliptic equations*, Proceedings of the Banach Center **33** (1996), 9–18.
4. H. Brézis, *Problèmes unilatéraux*, J. Math. Pure et Appl. **51** (1972), 1–162.
5. _____, *Opérateurs maximaux monotones et semigroupes de contractions dans les espaces de Hilbert*, Mathematics Studies, vol. 5, North Holland – New York, 1973.
6. _____, *Analyse fonctionnelle*, Masson, Paris, 1983.
7. M. Chipot, *Variational inequalities and flow in porous media*, Springer Verlag, New York, 1984.
8. _____, *Element of nonlinear analysis*, Birkhäuser Verlag, Basel, 2000.
9. M. Chipot and G. Michaille, *Uniqueness results and monotonicity properties for strongly nonlinear elliptic variational inequalities*, Ann. Scuola Norm. Sup. Pisa, Serie IV **16, 1** (1992), 137–166.
10. M. Chipot and J.F. Rodrigues, *On a class of nonlocal nonlinear elliptic problem*, M^2 AN **26, 3** (1992), 447–468.
11. M. Chipot and A. Rougirel, *On the asymptotic behaviour of the solution of elliptic problems in cylindrical domains becoming unbounded*, to appear.
12. _____, *On the asymptotic behaviour of the solution of parabolic problems in cylindrical domains of large size in some directions*, to appear.
13. _____, *Sur le comportement asymptotique de la solution de problèmes elliptiques dans des domaines cylindriques tendant vers l'infini*, CRAS Paris t.331, Série 1 (2000), 435–440.
14. P.G. Ciarlet, *The finite element method for elliptic problems*, North Holland, Amsterdam, 1978.
15. _____, *Elasticité tridimensionnelle*, Masson, 1986.
16. _____, *Elasticity*, North Holland, 1988.
17. R. Dautray and J.L. Lions, *Mathematical analysis and numerical methods for science and technology*, Springer Verlag, 1992.
18. A. Friedman, *Variational principles and free-boundary problems*, R.E. Krieger Publishing Company, Malabar, Florida, 1988.
19. D. Gilbarg and N.S. Trudinger, *Elliptic partial differential equations of second order*, Springer Verlag, 1983.
20. D. Kinderlehrer and G. Stampacchia, *An introduction to variational inequalities and their applications*, Academic Press, 1980.

21. O.A. Ladyženskaja, V.A. Solonikov, and N.N. Ural'ceva, *Linear and quasilinear equations of parabolic type translations of mathematical monographs*, vol. 23, AMS, 1968.

22. J.L. Lions, *Quelques méthodes de résolution des problèmes aux limites non linéaires*, Dunod–Gauthier-Villars, 1969.

23. J.L. Lions and G. Stampacchia, *Variational inequalities*, CPAM **20** (1967), 493–519.

24. C.B. Morey, *Multiple integrals in the calculus of variations*, Springer, 1966.

25. J. Nečas and I. Hlaváček, *Mathematical theory of elastic and elastico-plastic bodies: An introduction*, Studies in Applied Mechanics 3, Elsevier, 1981.

26. J.A. Nitsche, *On korn's second inequality, r.a.i.r.o.*, Analyse Numérique **15**, **3** (1981), 237–248.

27. A.S. Shamaev O.A. Olenik and G.A. Yosifian, *Mathematical problems in elasticity and homogenization. studies in mathematics and its applications*, vol. 26, 1992.

28. M.H. Protter and H.F. Weinberger, *Maximum principles in differential equations*, Springer Verlag, 1984.

29. P.A. Raviart and J.M. Thomas, *Introduction à l'analyse numérique des équations aux dérivées partielles*, Masson, 1983.

30. J.F. Rodrigues, *Obstacle problems in mathematical physics*, Math. Studies, vol. 134, North Holland, 1987.

31. A. Rougirel, *Private communication*.

32. E. Sandier and I. Shafrir, *On the symmetry of minimizing harmonic maps in n dimensions*, Differential and Integral Equations **6** (1993), 113–122.

33. L. Schwartz, *Théorie des distributions*, Hermann, 1966.

34. _____, *Topologie générale et analyse fonctionnelle*, Hermann, 1970.

35. G. Stampacchia, *Opere scelte*, vol. I, II, Edizioni Cremonese, 1997.

36. G.M. Troianiello, *Elliptic differential equations and obstacle problems*, Plenum, New York, 1987.

Index